HIGH-SPEED OPTICAL TRANSCEIVERS

Integrated Circuits Designs and Optical Devices Techniques

SELECTED TOPICS IN ELECTRONICS AND SYSTEMS

Editor-in-Chief: **M. S. Shur**

Published

Vol. 22: Topics in High Field Transport in Semiconductors
eds. *K. F. Brennan and P. P. Ruden*

Vol. 23: Oxide Reliability: A Summary of Silicon Oxide Wearout, Breakdown, and Reliability
ed. *D. J. Dumin*

Vol. 24: CMOS RF Modeling, Characterization and Applications
eds. *M. J. Deen and T. A. Fjeldly*

Vol. 25: Quantum Dots
eds. *E. Borovitskaya and M. S. Shur*

Vol. 26: Frontiers in Electronics: Future Chips
eds. *Y. S. Park, M. S. Shur and W. Tang*

Vol. 27: Intersubband Infrared Photodetectors
ed. *V. Ryzhii*

Vol. 28: Advanced Semiconductor Heterostructures: Novel Devices, Potential Device Applications and Basic Properties
eds. *M. Dutta and M. A. Stroscio*

Vol. 29: Compound Semiconductor Integrated Circuits
ed. *Tho T. Vu*

Vol. 30: Terahertz Sensing Technology — Vol. 1
Electronic Devices and Advanced Systems Technology
eds. *D. L. Woolard, W. R. Loerop and M. S. Shur*

Vol. 31: Advanced Device Modeling and Simulation
ed. *T. Grasser*

Vol. 32: Terahertz Sensing Technology — Vol. 2
Emerging Scientific Applications and Novel Device Concepts
eds. *D. L. Woolard, W. R. Loerop and M. S. Shur*

Vol. 33: GaN-Based Materials and Devices
eds. *M. S. Shur and R. F. Davis*

Vol. 34: Radiation Effects and Soft Errors in Integrated Circuits and Electronic Devices
eds. *R. D. Schrimpf and D. M. Fleetwood*

Vol. 35: Proceedings of the 2004 IEEE Lester Eastman Conference on High Performance Devices
ed. Robert E. Leoni III

Vol. 36: Breakdown Phenomena in Semiconductors and Semiconductor Devices
M. Levinshtein, J. Kostamovaara and S. Vainshtein

Vol. 37: Radiation Defect Engineering
Kozlovski V. and Abrosimova V.

Vol. 38 Design of High-Speed Communication Circuits
ed. R. Harjani

Selected Topics in Electronics and Systems – Vol. 39

HIGH-SPEED OPTICAL TRANSCEIVERS

Integrated Circuits Designs and Optical Devices Techniques

Yuyu Liu
Huazhong Yang
Tsinghua University, Beijing, China

World Scientific
NEW JERSEY • LONDON • SINGAPORE • BEIJING • SHANGHAI • HONG KONG • TAIPEI • CHENNAI

Published by

World Scientific Publishing Co. Pte. Ltd.
5 Toh Tuck Link, Singapore 596224
USA office: 27 Warren Street, Suite 401-402, Hackensack, NJ 07601
UK office: 57 Shelton Street, Covent Garden, London WC2H 9HE

British Library Cataloguing-in-Publication Data
A catalogue record for this book is available from the British Library.

HIGH-SPEED OPTICAL TRANSCEIVERS
Integrated Circuits Designs and Optical Devices Techniques

ISBN 981-256-641-4

Editor: Tjan Kwang Wei

Printed in Singapore by Mainland Press

PREFACE

With exploding data transmission traffic occurring every year, we are compelled to raise the bandwidth of communication networks. When this happens, the application of optical technologies does not have an alternative because of the unique transmission capacity or large bandwidth of optical fibers. Optical communication systems have been used since the 1970s for high-volume data transmission within wide-area, metropolitan-area and local-area networks, and its development has been rapid. The introduction of wavelength division multiplexing (WDM) technology increased the capacity of fiber by many folds and this quantum jump enabled a dramatic worldwide increase in Internet usage. Recently, the research target is directed towards higher spectral efficiency, since ultra dense wavelength-division-multiplexing (DWDM) transmission is the key to cost-effective capacity expansion in optical communication systems with a finite bandwidth. Today's accessible throughput of a single fiber is 2 Tb/s, which allows accommodating all telephone traffic of a planet through a single optical fiber. In laboratory conditions it was possible to reach the information transfer speed of 10 Tb/s, and the theoretical maximum is estimated about 300 Tb/s. In addition, the optical fiber has the following advantages: i) Low transmission loss (0.15 to 0.2dB/km), allowing longer repeater spacing and reducing outside plant expense. By comparison, the loss reaches 200 dB/km at 100 MHz for twisted-pair cables and 500 dB/km at 1 GHz for low-cost coaxial cables. Also wireless propagation with carrier frequencies of several giga-hertz incurs an attenuation of tens of decibels across a few meters while supporting data rates of lower than 100 Mb/s. ii) Absence of mutual influences between optical fibers (for example, no crosstalk). iii) Antijam capability without complex codes and technologies. Optical fibers are immune to electromagnetic interference and can be used in noisy environments.

The seemingly insatiable demand for bandwidth has led to the deployment of optical networks operating at several tens of gigabits per second. Commercial time-division multiplexed systems (TDM) running at 40Gbps are now a reality. Commercial deployment of ultra-high speed optical system depends on the availability of low cost and high performance optical/electronic devices and integrated circuits, which enable manufacturing to meet cost, performance, power and size constraints. However, in modern conditions when doubling of the traffic occurs every year, the demand for communication capacity has outpaced the performance of semiconductor systems, which has doubled every 18 months according to Moore's law. On the other hand, the high inherent bandwidth of optical fibers is just beginning to be exploited. Thus long-term, high-risk and hard-going research is necessary to sustain the growth in information transfer capacity.

A basic optical communication system consists of six parts: (1) a serializer, which converts many low-speed parallel data streams into a high-speed serial data stream; (2) electro-optical transducer (e.g., a laser diode) and the relevant driver, or a laser source and a modulator; (3) a fiber; (4) a photodetector (e.g., photodiode); (5) amplifiers including a transimpedance amplifier (TIA) and a limiting amplifier, which amplifies the photodiode output with low noise and sufficient bandwidth, converting it to a large enough voltage signal; (6) a deserializer. The first thee parts makes a transmitter and the last three parts makes a receiver.

The transmitter entails several issues that manifest themselves at high-speeds and in scaled IC technologies. Since the jitter of the transmitted data is determined primarily by that of the frequency synthesizer, a robust, low-noise design with high antijam capability becomes essential. Furthermore, the design of skew-free multiplexers becomes difficult at high data rates. Another critical challenge arises from the laser driver, which is a circuit that must deliver tens of milli-amperes of current with very short rise/fall times and large voltage swings. Since optical modulator in the transmitting end needs to be driven with a large differential waveform, the modulator driver design becomes more difficult as scaled technologies impose lower supply voltages. The optical components of an optical transmission system, such as the laser diode, the fiber, and the photodiode, introduce their own nonidealities and these defects need close interaction between electronic and optical design. Phenomena and guidelines, such as chirp, dispersion, attenuation and efficiency, play a major role in the overall link budget. The receiver in an optical transmission system also suffers from many problems. The noise, gain, and bandwidth of the TIA and the limiting amplifier directly impact both the sensitivity and the speed of the overall system and these specifications will present additional problems as the supply voltage scales down. Moreover, the clock and data recovery (CDR) circuits must satisfy severe jitter and bandwidth requirements and provide a capability of tolerance of long runs, which is a sequence of identical bits.

Full integration of the transceiver on a single chip is another challenge raising a number of concerns. The area saving coming from full integration could be at the expense of circuit performance. For example, when analog front-end circuits are integrated with high-speed digital parts, such as multiplexer and de multiplexer in a low cost CMOS process, the performance of the sensitive circuits will suffer from high switching noise through substrate coupling. The above issues have resulted in multi-chip solutions that integrate the noisy and sensitive functions on different substrates. Recent work has integrated the serializer and deserializer (namely a "SERDES") in a single chip, but the TX and RX amplifiers may remain in isolation. At high speeds and/or high port densities, the power dissipation of optical transceivers becomes critical as it determines the type and size of the package in which the entire module is housed. Integrating a complete SERDES on a single CMOS chip serves as the first step toward much greater intricacy in optical communication transceiver design, and there are still considerable challenges for designers to achieve full integration. In particular, when it comes to full integration, power and area reduction, and noise isolation are the hot topics.

On the whole, increasing the transmission capacity of optical communication systems will require many breakthroughs in research. Novel tunable lasers and amplifiers, high-speed modulators operating between 1200nm-1700nm will need to be devised to open up wider fiber bandwidth. Adaptive devices such as filters, gain and power controllers, and dispersion compensators become increasingly important, as they must be used to overcome the nonideal effects which are no longer possible to neglect at ultra-high transmission speed. Achieving full integration and packaging remain a challenge for broadband optical communications systems with the increasing complexity, scale and further demand for low-cost, high-volume manufacturing, which is crucial to meet the requirement of last-mile applications. These expected targets are obtained as higher level of integration is achieved and new design ideology and fabrication technology are devised.

The design of high-speed optical devices and integrated circuits for optical transceivers and links oriented to low power, low cost and small area will be discussed in this special issue. In addition, other relevant technologies of optical transceivers will also be introduced.

Laser is a key component for optical communication systems. Many techniques, such as soliton dispersion management techniques for novel transmission systems at 40 Gb/s/channel and higher bit rates, rely on clean optical pulses and thus benefit greatly from the availability of simple, compact, low-noise and efficient optical pulse generating lasers. Lasers are becoming increasingly important for telecommunication applications as data transmission rates continue to increase. Gabriel Spuehler and his coworkers reported their new efforts to develope various novel multi-GHz laser sources and their application in optical communication field, such as in the optical time-division multiplexing (OTDM) and DWDM systems. Optical modulator in the transmitting end needs to be driven with a large differential waveform. The design of the modulator driver presents some unique challenge due to the high speeds and high voltage swings involved. Shanthi Pavan et al. discussed design considerations for high-speed high-swing integrated differential modulator drivers in SiGe technology.

Due to the cost consideration, it is a challenge to effectively make use of the huge installed fiber in higher speed optical communication systems without degrading the repeater spacing and transport span between the nodes. This presents a significant challenge because 10Gb/s signals are much more susceptible to the impairments introduced by the installed fiber, such as inter symbol interference (ISI). ISI is a fundamental problem in a digital communication system such as Ethernets and optical backplanes. For a given data rate, the ISI limits the achievable link distance without the optical repeater, due to the fact that the ISI increases exponentially with the transmission distance. Any further increase in the data rate would imply shorter distances. Then, there is a trade off between the data rate and the transmission distance without optical repeater. The main source of ISI is signal pulse broadening due to fiber dispersion. In order to mitigate the effects of dispersion, different methods have been proposed and implemented in the three parts of an optical link span, i.e. transmitter, fiber and receiver, and in both optical and electrical domain. Although the compensation methods can be achieved in optical domain, electrical schemes, in particular equalizer, have attracted a more considerable amount of interest because they allow greater integration with existing circuitry, leading to more compact, less expensive and more adaptive solutions. This is especially true for WDM systems, in which every channel needs dispersion compensation. Several techniques to alleviate dispersion are introduced in this special issue. Prof. Manfred Berroth and Rui Tao designed a monolithically integrated duobinary optical transmitter, which adopts duobinary coding technology as a dispersion compensation method due to its higher spectral efficiency. A 10Gb/s analog continuous-time equalizer with integrated clock and data recovery circuit was presented by Douglas S. McPherson and coworkers and it can be used to recover signals degraded by chromatic and polarization mode dispersion effectively. Jonathan Sewter and Anthony Chan Carusone give a unique and detailed analysis of polarization-mode dispersion (PMD) effects in a 40-Gb/s optical system and the analysis results are used to compare different electronic equalizer architectures as potential PMD compensators. Hence, an increase in maximum transmission length of more than several times or in higher system transmission capacity is achieved.

Serial inputs and outputs (IOs) are widely used in optical communication systems to achieve high-speed data transmission. A key performance for high-speed serial IOs in optical backplane transceivers is the jitter characteristic. Clock and data recovery (CDR) circuit is a crucial component for the jitter characteristic of an optical serial transceiver. Hence, CDR circuits must satisfy stringent specifications defined by optical standards and this presents challenges to system and circuit designers. Although several ways exist to perform clock recovery, for example based on non-linear processing and surface acoustic wave (SAW) filters, monolithic implementations are usually based on a phase locked loop (PLL). Shenggao Li provided an overview of high-speed serial IO design trade-offs over a wide scope ranging from system partitioning, architecture selection, to circuit design techniques, with emphasis on jitter reduction, power and cost saving. Cicero S. Vaucher et al. gave a unique and detailed engineering presentation on the subject of high performance architectures and building blocks for clock and data recovery (CDR) applications and introduces a parallel CDR architecture for high operation speeds and low-power dissipation.

A novel and useful logic style for high-speed, low-power and mixed-signal integrated circuits applications was presented by Yuyu Liu and her coworkers, which is called MOS Current Mode Logic (MCML). MCML circuits seem to be very increasingly popular in the design field of optical transcerivers, such as in designs for gigahertz multiplexer/demultiplexer and CDR circuits.

Key front-end ICs have been demonstrated in the past using HBT (SiGe, InP, GaAs) and HEMT (InP, GaAs) technologies. Among them InP double hetero-junction bipolar transistor (DHBT) technology exhibits excellent high-speed performance, high breakdown voltage, large-scale integration, good uniformity and reliability. This makes them suitable for realizing both broadband analog amplifiers and large-scale digital circuits required for the optical transceiver components in 40Gb/s or higher speed systems. Kiyoshi Ishii et al. presented their InP-based heterojunction bipolar transistor (HBT) technologies and relevant circuit design techniques, attractive for fabricating ultrahigh-speed integrated circuits for 40 to 100 Gbit/s-class optical communications systems.

From the point of the whole system, a TDM system transceiver with 4–channel 10 Gb/s interfaces to achieve 40 Gb/s non-return to zero (NRZ) optical transmission was presented by Karthikeyan Krishnamurthy et al. Haruhiko Ichino and his coworkers implemented an XENPAK optical transceiver with inter-frame ling signaling (ILS) interface technology to provide a new seamless connection technology between an optical transport network, such as SDH/SONET, and a 10-Gigabit Ethernet. The ILS interface technology is expected to play an important role in achieving highly reliable and cost-effective Ethernet native optical WANs in the future.

In conclusion, new optical communication technologies catering to various system requirements continue to shrive, with the ultimate goal of cramming more information onto relatively limited channels of limited bandwidth using minimum power.

The editors would like to express great gratitude to all the authors for their outstanding contributions to this special issue of "SPECIAL ISSUE ON HIGH-SPEED OPTICAL TRANSCEIVERS: INTEGRATED CIRCUITS DESIGN & OPTICAL DEVICES TECHNIQUE". The Special Issue will prove useful for a large audience: from practicing engineers and scientists to graduate students, who are interested in the development of integrated circuits and devices technology for optical communication applications.

Yuyu Liu* and Huazhong Yang† (Guest Editors)
Department of Electronic Engineering,
Tsinghua University,
Beijing 100084, China
*liu-yy02@mails.tsinghua.edu.cn
†yanghz@tsinghua.edu.cn

CONTENTS

International Journal of High Speed Electronics and Systems
Vol. 15, No. 3 (2005) 477–495

World Scientific
www.worldscientific.com

DESIGN CONSIDERATIONS FOR INTEGRATED MODULATOR DRIVERS IN SILICON GERMANIUM TECHNOLOGY

SHANTHI PAVAN[†]

Department of Electrical Engineering
Indian Institute of Technology Madras
Chennai 600 036, India
[†]shanthi@ee.iitm.ac.in

MAURICE TARSIA[§], STEFFEN KUDSZUS, DAVID PRITZKAU

Bigbear Networks, 345 Potrero Avenue
Sunnyvale, CA 94085, USA
[§]tarsia@bigbearnetworks.com

We present design considerations for high speed high swing differential modulator drivers in SiGe BiCMOS technology. Trade-offs between lumped and distributed designs, and linear and limiting amplifiers are examined. The design of a 6 V output modulator driver is discussed in detail. The driver features a unique bias generation and distribution circuit that enables low power-supply operation. Simulation results and measurements are given.

Keywords: modulator;driver;distributed;amplifier;limiting;bias;SiGe.

1. Introduction

The seemingly insatiable demand for bandwidth has led to the deployment of optical networks operating at several tens of gigabits per second. Commercial time division multiplexed systems (TDM) running at 40 Gbps are now a reality.[1 2 3] The block diagram of a typical OC-768 transceiver is shown in Figure 1. Four 10 Gbps datastreams are multiplexed together. This stream, with a data rate of 40 Gbps, is amplified and drives a Mach-Zehnder interferometric optical modulator. The optical modulator needs to be driven with a large differential waveform.

At the receiver, a photodiode is used to convert the optical signal into an electrical one. The photodiode current is amplified by a transimpedance amplifier (TIA) and further processed by a limiting amplifier, which converts the small signal TIA output to a large two-level signal. This in turn drivers the clock and data recovery unit (CDR), and a demultiplexer that splits the 40 Gbps stream into four 10 Gbps datastreams. While much of the electronics in earlier generation OC-768 transponders was designed in compound semiconductor technologies,[4 5 6] it has now become feasible to realize most of the functionality in silicon germanium (SiGe) BiCMOS processes. SiGe provides an opportunity for unparalleled system integration, as high

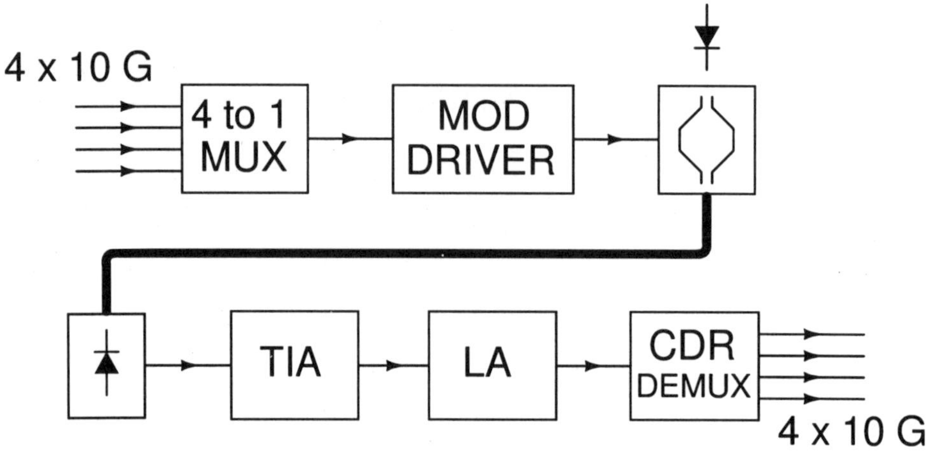

Fig. 1. Block diagram of an OC-768 transceiver.

performance bipolar transistors and CMOS devices can be included on the same chip. Many of the building blocks for OC-768 transceivers, like multiplexers and demultiplexers (MUX/DEMUX), transimpedance amplifiers (TIA), limiting amplifiers (LA) and clock and data recovery units (CDR) have been reported in SiGe BiCMOS processes.[2 3] These are testimony to the maturity and reliability of state of the art SiGe processes.

The design of the modulator driver presents some unique challenges due to the high speeds and high voltage swings involved. [8 9 10] A typical value for the common-emitter breakdown voltage, $BV_{CE,O}$ in these technologies is about 1.8 V. Other major challenges are electro-migration in the interconnect metalization, and the formation of hot-spots due to oxide isolation of the bipolar transistors. Hot-spots could degrade device performance due to self-heating. In order to reduce the power dissipation of the driver, it is necessary to keep the power supply as small as possible. In this work, we present the design of a modulator driver with a fully differential output swing of 6 V peak-to-peak swing, implemented in a SiGe process, with nominal supply voltage of 6 V. Designed for operation at OC-768 data rates, the driver features a unique bias distribution technique. The rest of this paper is organized as follows. In Section 2, we compare the merits of a lumped versus distributed driver design and present the evolution of the gaincell used in the design. In Section 3 we show the chip architecture and discuss circuit details. We present in detail the development of the tail current bias control technique in Section 4. We show simulations and experimental results in Sections 5 and 6 respectively. Conclusions are given in Section 7.

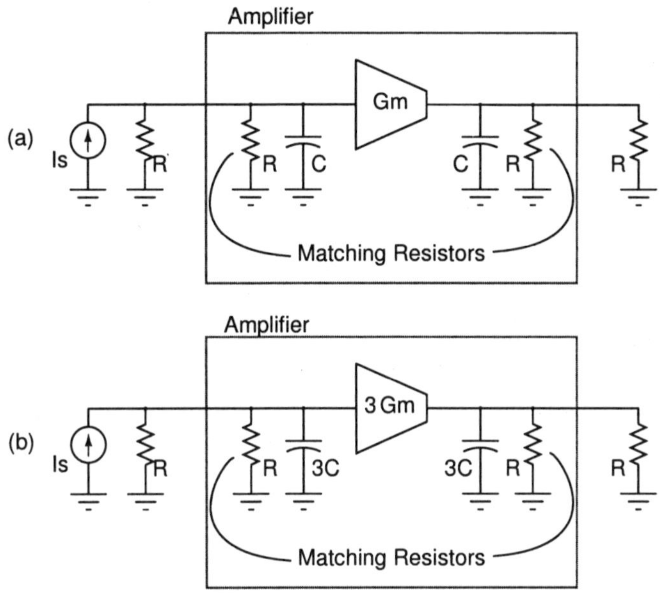

Fig. 2. Gain bandwidth limitations of lumped ampifiers.

2. Driver Architecture

2.1. *Lumped versus Distributed Design*

Consider the design of an amplifier using a transconductor (G_m) with input and output capacitors of value C, as shown in Figure 2(a). The source and load impedances are resistors of value R. At low frequencies, the input and output impedances of the amplifier are matched to the source and load respectively. The input and output capacitances and the resistors form poles that result in a finite bandwidth, which is inversely proportional to RC. The low frequency gain is proportional to $G_m R$, so the gain-bandwidth product of the amplifier is proportional to G_m/C.

In order to obtain a higher DC gain (e.g $3\,G_m\,R$), the transconductance has to be increased by a factor of three, as shown in Figure 2(b). This increase in gain is accompanied by a proportional decrease in bandwidth due to the three-fold increase in input and output parasitic capacitances. Hence, the gain-bandwidth product remains the same as in the low-gain case. A technique to increase the gain-bandwidth product is distributed amplification, where the input and output capacitances of the active device are "absorbed" into a synthetic transmission line, as shown in Figure 3. The characteristic impedance of this line is $Z_o = \sqrt{\frac{L}{C}}$. Thus, L must be chosen to be equal to R^2C. For a thorough exposition of distributed amplifiers, the reader is referred to Wong's book [11]. It can be shown that a distributed amplifier trades off bandwidth for latency, and the performance depends critically on the

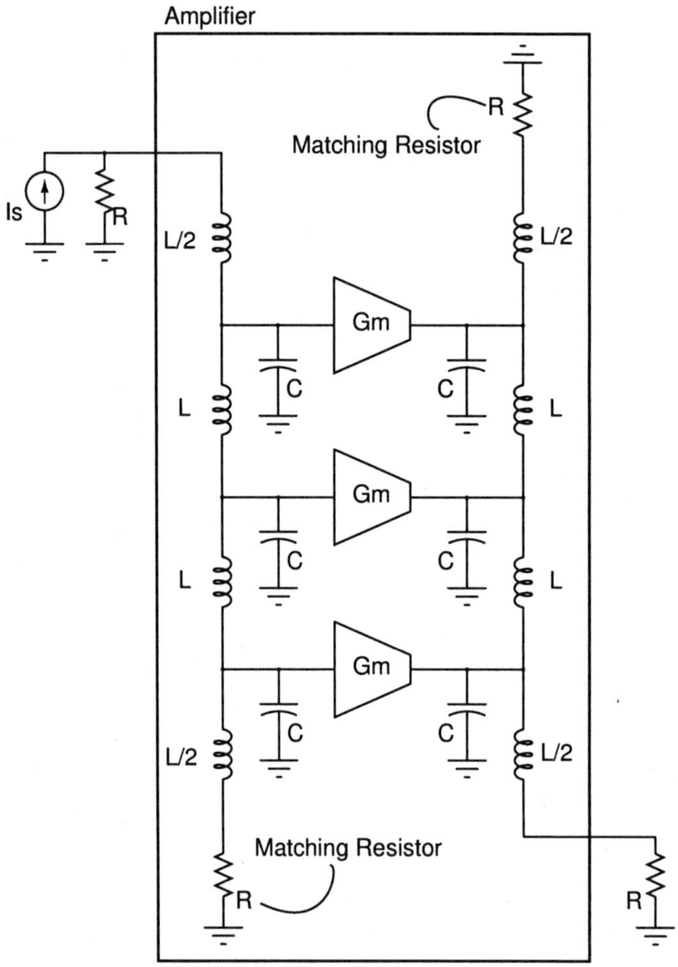

Fig. 3. The distributed amplifier with lumped elements.

delay synchronization between the input and output transmission lines. While the amplifier in Figure 3 uses lumped inductors for broad banding, the same effect can be achieved with transmission lines, as shown in Figure 4. τ represents the time of flight of the transmission lines. Noticing that a transmission line section can be approximated as shown toward the right of the figure, we see that the value of Z_1 and τ need to satisfy

$$R = \sqrt{\frac{Z_1 \tau}{\frac{\tau}{Z_1} + C}} \tag{1}$$

Given Z_1, R and C, the delay of the transmission line required can be calculated

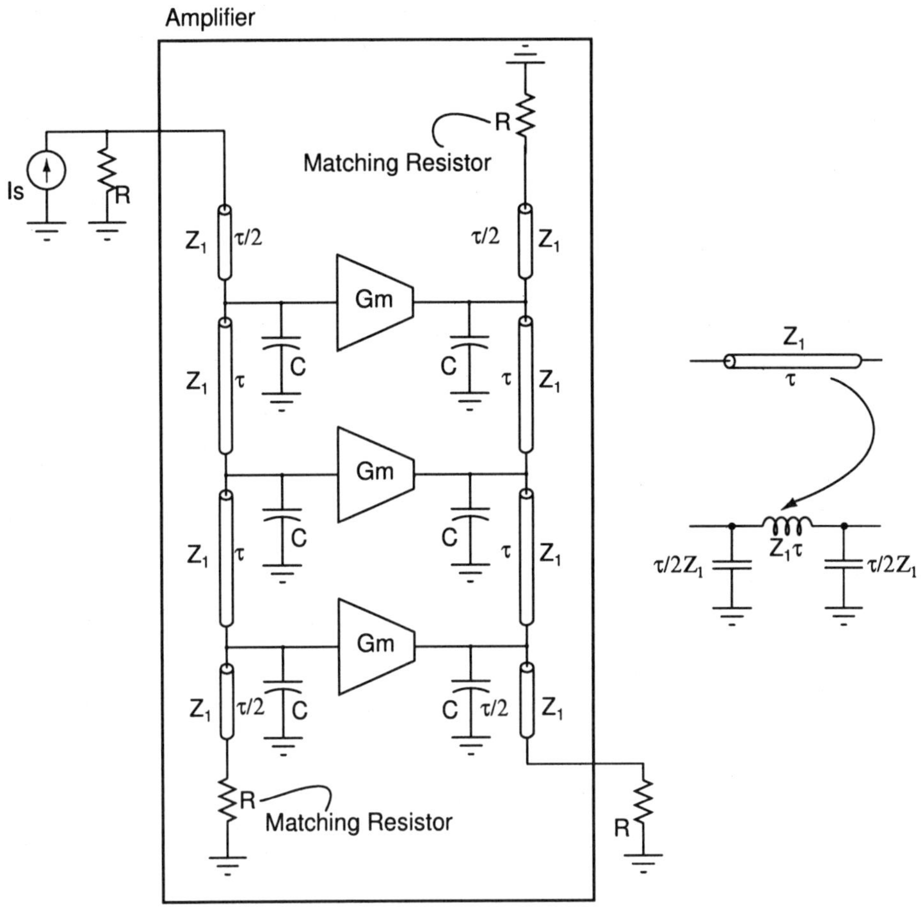

Fig. 4. The distributed amplifier with transmission line broadbanding.

to be

$$\tau = CR\frac{Z_1/R}{(Z_1/R)^2 - 1} \tag{2}$$

A microstrip line was chosen as the transmission line structure in our chip. The characteristic impedance of a microstrip line decreases with increasing width, so from the discussion above, we would like to make the line narrow. In this application, however, the output lines are required to carry large amounts of current, which necessitates a wide line. Hence, large voltage swings and high bandwidth have inherently contradictory requirements.

In our design the input capacitance of the "transconductors" was about 30 fF. A characteristic impedance of about 65 Ω was chosen for Z_1 as a compromise between chip dimensions and electromigration in the output transmission lines.

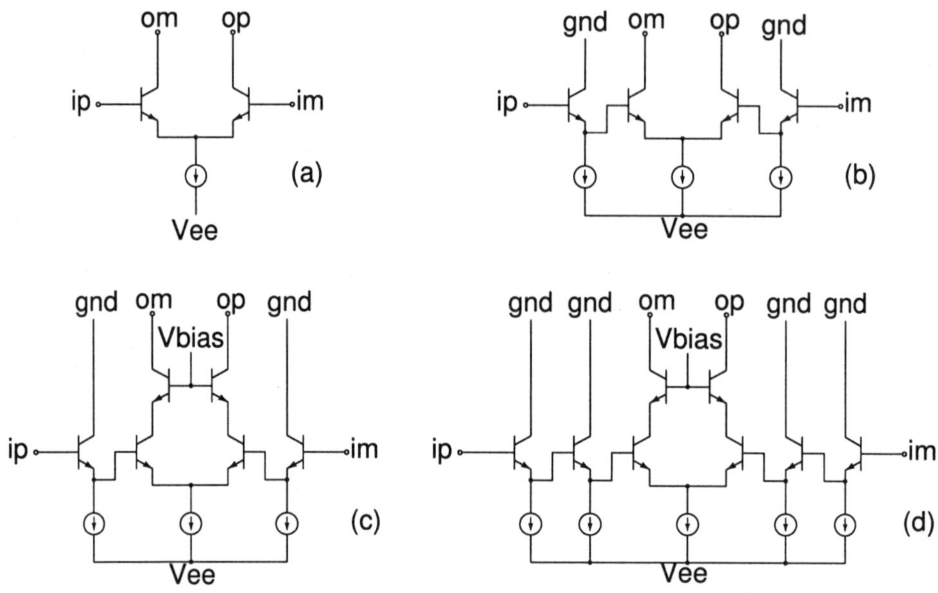

Fig. 5. Evolution of the gaincell.

2.2. *Linear versus Limiting Gaincell*

Another key design decision that needs to be made is regarding the input-output behavior of the gaincell - should this be linear, or could it be a limiting type of design ? It turns out that the answer to this question depends on the application. This design being intended for NRZ data, linearity is not required. Moreover, a linear gaincell would consume a lot more power than a limiting one because only a small portion of the quiescent current would be modulated by the input signal. Since power dissipation of the entire driver is of significant concern, a limiting gaincell is one of choice here.

2.3. *Fully Differential Gain Cell Design*

Figure 5 shows the evolution of the gaincell. The very basic current switch that come to mind is shown as (a) in the figure. Here the tail current is switched between op and om depending on the sign of the input differential voltage. It is being easy to layout, but the input and output capacitances are vastly different. Moreover, the input capacitance is lossy. These reasons make it difficult to synchronize the delays of the input and output lines of the traveling wave structure. The overlap capacitances between input and output nodes also cause problems.

The structure of Figure 5(b) uses emitter-followers to drive the input of the current-switching pair. This bring the input and output capacitances closer, but still suffers from the problems caused by input-output overlap capacitances of the

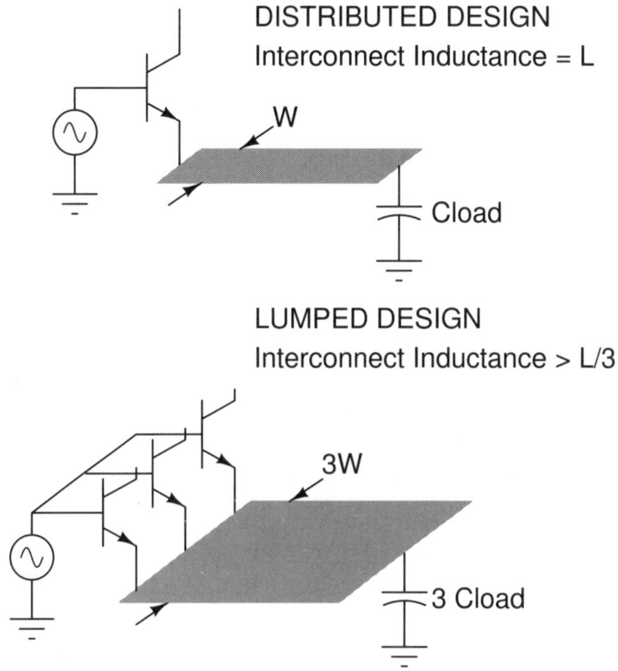

Fig. 6. Lumped versus distributed gaincells with interconnect inductance.

switch.

Input-output isolation is improved by using a common-base stage at the output of the current switch as shown in Figure 5(c). We found that another buffer stage was required to make the input and output capacitances of the gaincell equal. The final topology of the gaincell is shown in Figure 5(d).

It is apt to mention at this juncture some additional benefits that accrue from the distributed nature of the driver. Notice that the gaincell has several emitter followers driving successive stages, that typicallly present a capacitive load. Due to the large bias currents involved in the design, the impedance levels of operation are extremely low (few ohms). This means that inductance of the interconnect can severely degrade the performance of the circuit. Consider the case of an emitter follower driving a capacitive load in a 3 stage distributed amplifier, as shown in Figure 6. The width of the wire connecting the emitter to the capacitive load is denoted by W, while its series inductance is denoted by L. Assume that the design is stable. Consider now a lumped version, shown toward the lower part of the figure. It has to drive thrice the load capacitance with one-third the source impedance. It would be just as stable as the distributed amplifier if the interconnect inductance scaled to $L/3$. However, the inductance is much higher than $L/3$ (even though the wire width is 3W) due to mutual coupling effects. Hence, the stability of the lumped

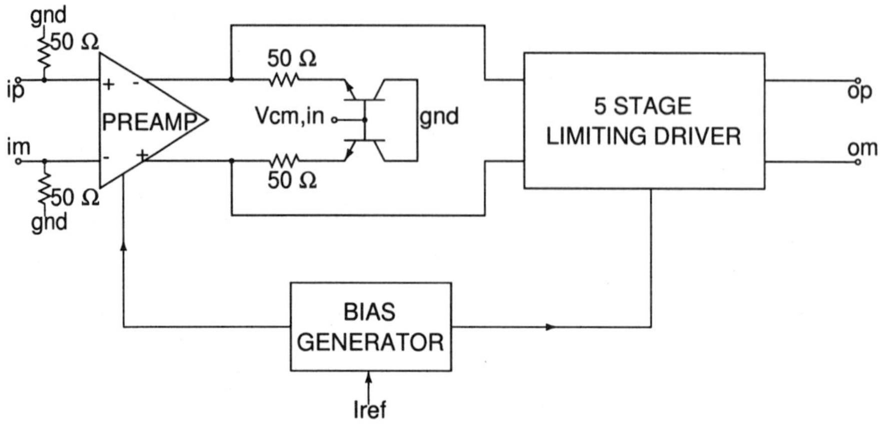

Fig. 7. Modulator driver chip architecture.

version is degraded.

Another aspect where a distributed design wins over a lumped one is that of thermal hot-spots. In a lumped design all the power is dissipated in one area, potentially heating up local interconnect and reducing long term reliability. On the other hand, a distributed design has many smaller sources of heat, distributed over the chip, which is thermally a more favored situation.

3. Circuit Details

The architecture of the entire chip is shown in Figure 7. It consists of a preamplifier driving a five stage distributed limiting amplifier. The preamplifier (shown in Figure 8) takes in an external input and generates a fully differential $600\,\mathrm{mV_{pp}}$ drive at a correct common-mode voltage suitable for the main driver. The first emitter follower level shifts the input common mode, while the second is used to drive the current switch.

For the driver, a five stage design represented a reasonable trade-off between the cut-off frequency of the artificial transmission input and output transmission lines on one hand and complexity involved in biasing multiple gaincells on the other. The power supply voltage was chosen to be $V_{ee} = -6\,\mathrm{V}$.

The schematic of the gaincell used in the main amplifier is shown in Figure 9. The emitter followers were biased with resistors and diodes. Their collectors are connected to a voltage lower than ground, to prevent breakdown of the devices. The tail current source of the current switching pair is biased with a MOSFET in order to save headroom. The low output resistance of the MOSFET, combined with the desired stability of the driver peak-to-peak output swing necessitates very careful biasing of the tail current source. This is discussed in detail in the next section.

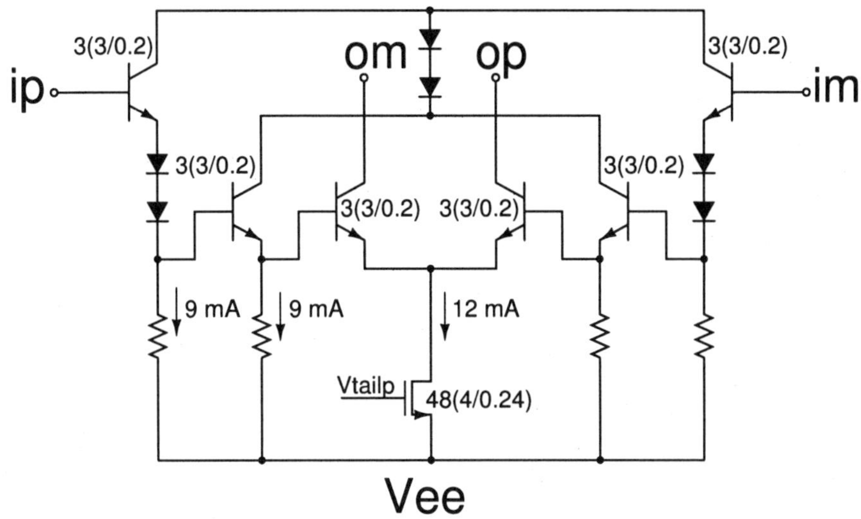

Fig. 8. Schematic of the preamplifier.

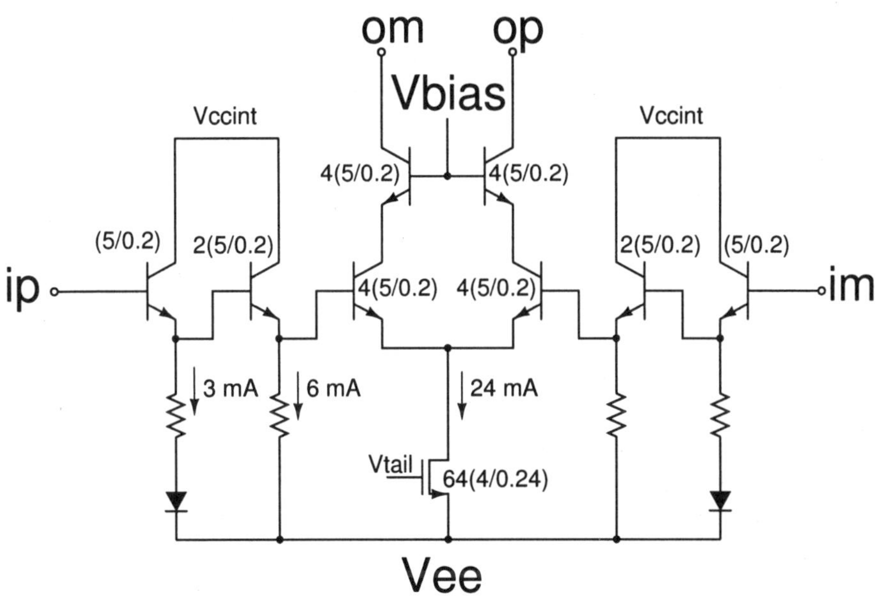

Fig. 9. Schematic of final gaincell used in the main amplifier.

4. Tail Current Bias Generation & Distribution

In this section, we discuss the important issue of accurately generating the required tail current for the current switching pair. Since the driver is a limiting distributed

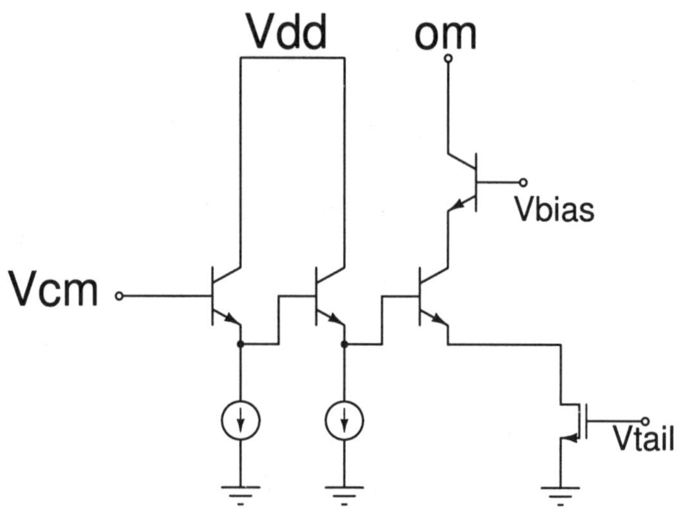

Fig. 10. Half circuit of the gaincell.

amplifier design, the peak-to-peak output swing is directly proportional to the tail current - so this current must be constant across power supply voltage, process variations and temperature. Further, as we wish to minimize the power supply voltage required for the driver IC, a MOS transistor is a more appropriate choice for realizing the current source. The half-circuit of the gaincell incorporating a MOS tail current source is shown in Figure 10.

We are faced with two basic choices for generating and distributing Vtail.

- Global Vtail generation : Here the required bias voltage is generated at one point on the chip and distributed to all transistors that require it, as shown in Figure 11. This approach has the advantage that very little extra power is required by the bias generator. However, there are several serious drawbacks associated with this scheme. Due to the distributed topology and high currents involved in the design there could be significant voltage drops along the ground bus (i.e, points a,b,c are not at the same potential). This means that the actual gate-source voltages appearing across M1-M5 could be very different from what was intended. Any threshold voltage mismatch between the devices used in the bias generation block and the actual current source transistors would also lead to tail currents that were poorly controlled over process and temperature variations. Hence, global generation of Vtail is not a viable approach.
- Local Vtail generation : In this approach, a bias generator is placed in proximity to each gain cell, as shown in Figure 12. The reference current Iref has to now be distributed to each gaincell, so the total power dissipation in the bias section increases. However, since the bias generator is placed

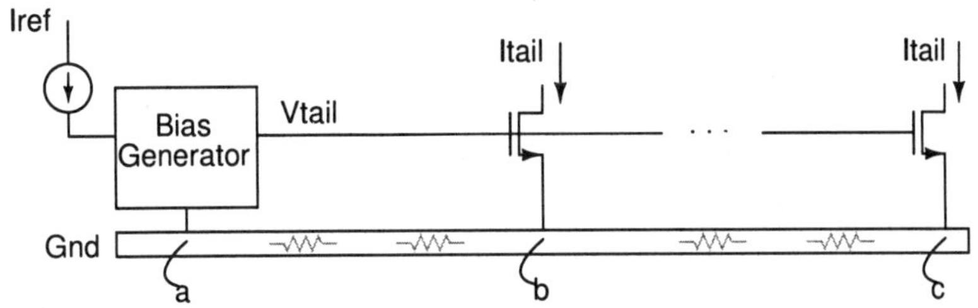

Fig. 11. Global Vtail generation & distribution.

Fig. 12. Local Vtail generation.

close to the gaincell, ground drops between gaincells is of no consequence. Further, this technique only demands MOS transistor threshold voltage matching within a gaincell, in contrast to a global Vtail generator, where matching must be maintained across great distances on the die. Hence, the local bias generation approach is used in this work.

We now present the operating principle of a precision bias circuit used in this work. The schematic is shown in Figure 13, where the gaincell is shown towards the right. M1', D1, D2 and Q3' in the bias generation network are sized so that they have the same current density as M1,Q1,Q2 and Q3 respectively. A current Iref is forced into the collector of Q3' and the negative feedback loop formed through Q6, D5, D4 and M2' determines Vtail. Note that M1' and M1 operate with identical terminal voltages regardless of process variations or temperature, so that in the absence of mismatch, the tail current is determined given by $\mathrm{Itail} = \mathrm{n\,Iref}$. Cc is the compensating capacitor for the bias negative feedback loop. The DC β's of the bipolar transistors in SiGe are very high (several hundreds), so there is negligible error due to the base current of Q6.

While this replica feedback bias generator is very precise, it is complicated to lay out. Note that Vcm and Iref lines need to be propagated to all the gaincells. With this, there also exists the possibility of parasitic coupling between gaincells through the Vcm bias line. The capacitor Cc is of the order of several picofarads, and is therefore unwieldy to place in the gain cell. As far as ease of layout is concerned, the most desirable circuit is shown toward the right of Figure 14. Here, only one

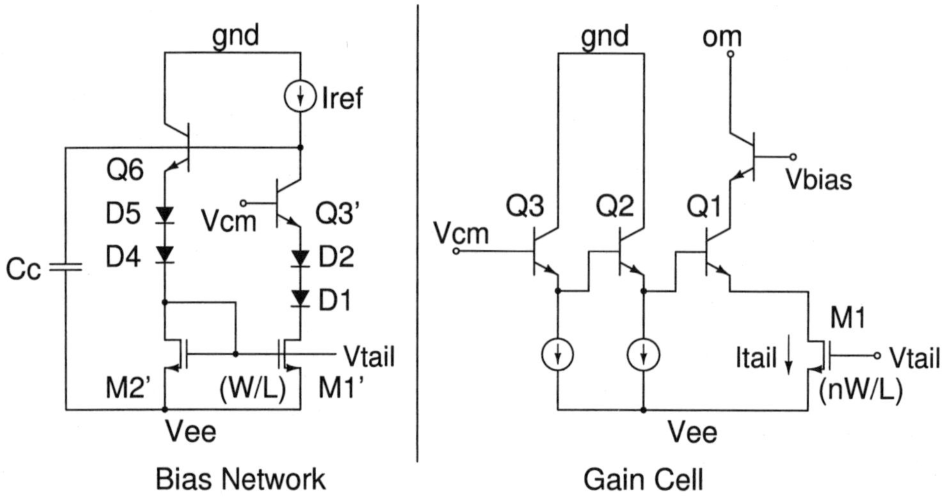

Fig. 13. Precision bias voltage generation using replica feedback.

line (Iref) goes to every cell, and no compensation capacitor is needed. However, it has the disadvantage that the tail current in the gaincell is not precisely set due to the difference in drain-source voltage of the FETs in the bias generator and the gaincell. The tail current can be shown to be

$$\mathrm{Itail} = \mathrm{n\,Iref}\frac{1+\lambda \mathrm{V_{DS,M1}}}{1+\lambda \mathrm{V_{DS,M1'}}} \tag{3}$$

where λ is the channel length modulation parameter of the MOSFETs. Itail is not precise since λ, $\mathrm{V_{DS,M1}}$and$\mathrm{V_{DS,M1'}}$ vary with process and temperature.

An improved bias circuit which combines the advantages of both techniques discussed above is possible. The kernel of the idea is the following - if the reference current to the layout friendly circuit is precompensated for the difference in V_{DS} of the FETs in the bias generator and the tail current source, then good precision can be obtained. The implementation of this principle is shown is shown in Figure 15. Ip, the predistorted current in M2' is seen to be

$$\mathrm{Ip} = \mathrm{Iref}\,\frac{1+\lambda \mathrm{V_{DS,M2'}}}{1+\lambda \mathrm{V_{DS,M1a'}}} \tag{4}$$

Hence the tail current is given by

$$\mathrm{Itail} = \mathrm{n\,Iref}\,\frac{1+\lambda \mathrm{V_{DS,M2'}}}{1+\lambda \mathrm{V_{DS,M1a'}}}\,\frac{1+\lambda \mathrm{V_{DS,M1}}}{1+\lambda \mathrm{V_{DS,M2a'}}} \tag{5}$$

Since $\mathrm{V_{DS,M2'}} = \mathrm{V_{DS,M2'}}$ and $\mathrm{V_{DS,M1a'}} = \mathrm{V_{DS,M1}}$, we see that $\mathrm{Itail} = \mathrm{n\,Iref}$ and is independent of process and temperature. The predistortion block is common to all

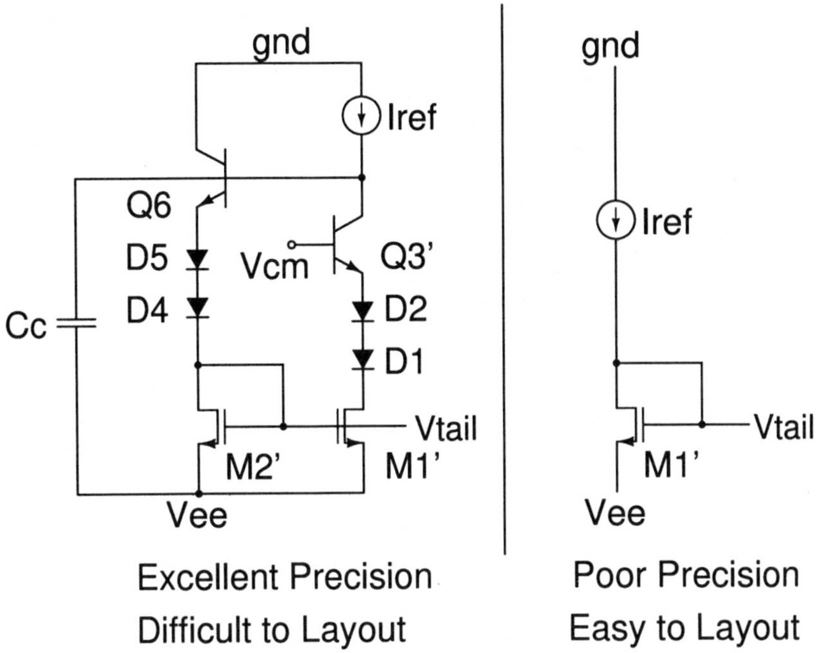

Fig. 14. Comparison of two bias generation techniques.

the gaincells and is placed at one corner of the chip. The predistorted currents are propagated to the gain cells. Layout is straightforward as the bias generator is a simple structure. Since currents are distributed, there is no problem wit regard to drops along the ground lines as explained earlier in this section.

5. Simulated Performance

In this section, we present simulated results for the modulator driver IC. Figure 16 shows the drive voltages at the inputs of the gaincells of the main amplifier. Notice that the peak-to-peak differential drive voltage is 600 mV (which is the output drive of the preamplifier) and the delay in the drives to successive stages is about 3.2 picoseconds. Figure 17 shows the driver output with an input step. It shows a 6 V peak-to-peak differential output voltage with a 10-90 % risetime of 10.8 ps. Figure 18 shows the simulated eye-diagram. A summary of thew simulated performance is shown in Table 1.

6. Experimental Results

In this section we present measurement results from a modulator driver fabricated in a 0.18 μm SiGe BiCMOS technology. The chip micro photograph is shown in Figure 19. The chip area is about 1.5 mm^2. Figure 20 shows the eye-diagram of the

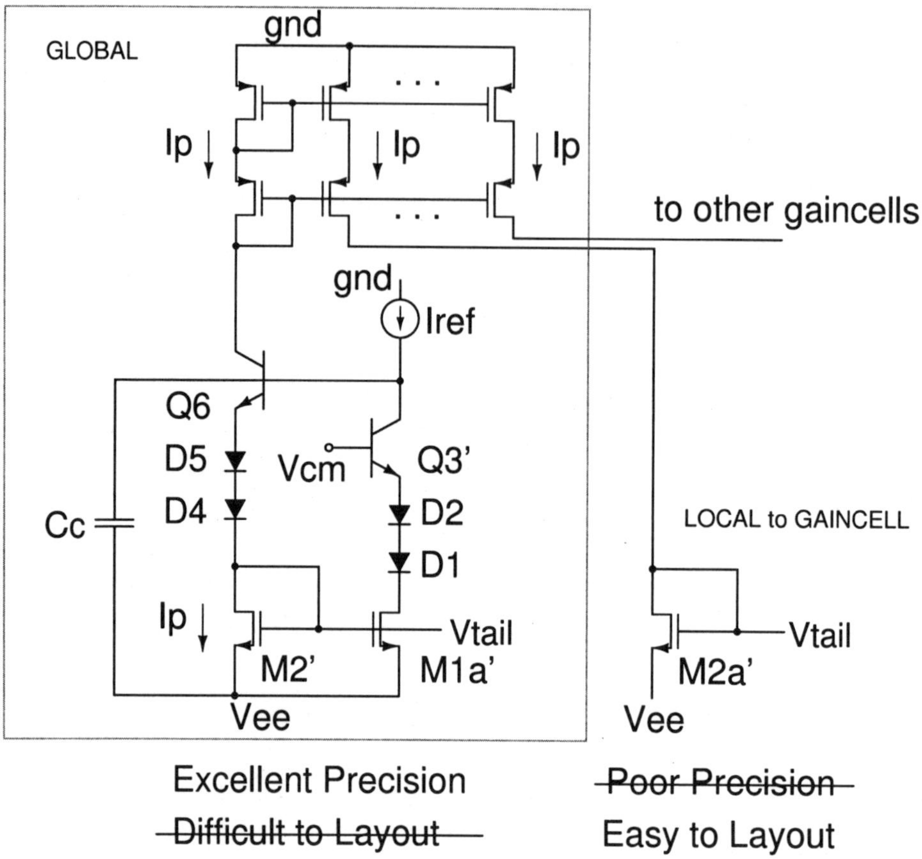

Fig. 15. Precision bias technique with simplified layout.

Table 1. Summary of simulated characteristics (25°C)

Technology	0.18 μm SiGe BiCMOS
Supply voltage	-6.0 V
Output Voltage Swing	6 V (peak-to-peak differential)
Input Voltage Drive	400-600 mV (peak-to-peak differential)
Rise Time (20-80%)	6.7 ps
Rise Time (10-90%)	10.8 ps
Chip area	1.5 mm^2
Current Drain from Vee	$\approx$ 300 mA
DC gain	20 dB

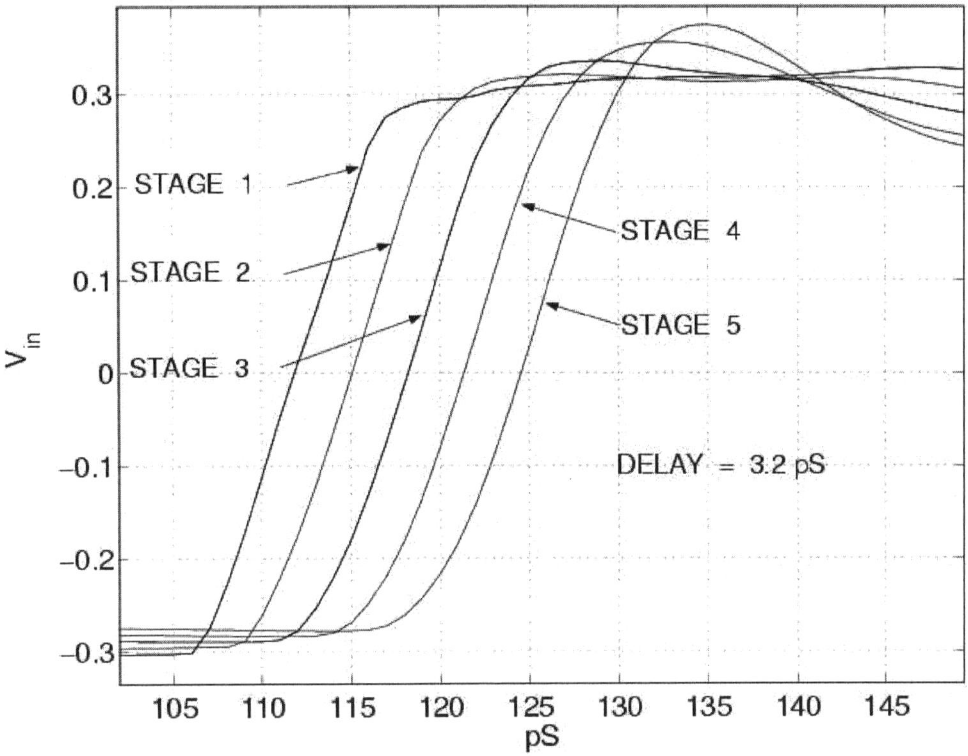

Fig. 16. Inputs to the successive gaincells of the modulator driver.

driver at a data rate of 12.5 Gbps. The input was provided by an Anristu MP1758A pulse-pattern generator. The output eyes were recorded with an Agilent 86100A DCA using the Agilent 83484A 50GHz plug-in module. A 16dB attenuator was used on the output of the modulator driver before measuring it on the DCA. The 12.5 GHz clock output of the pattern generator was used to trigger the DCA. To generate a 40 Gbps stream four 10 Gbps streams from the Anritsu pulse generator were multiplexed using an in-house 4-to-1 mux. The resulting single-ended driver output eye is shown in Figure 21. Eye closure is attributed to the test setup and the quality of the input eye to the driver.

7. Conclusions

Design considerations for high-speed high swing modulator drivers in SiGe technology were presented. A novel technique for generating and distributing bias voltages was discussed. Simulation and experimental results for a 6 V_{pp} differential modulator driver designed in a 0.18 μm technology were given.

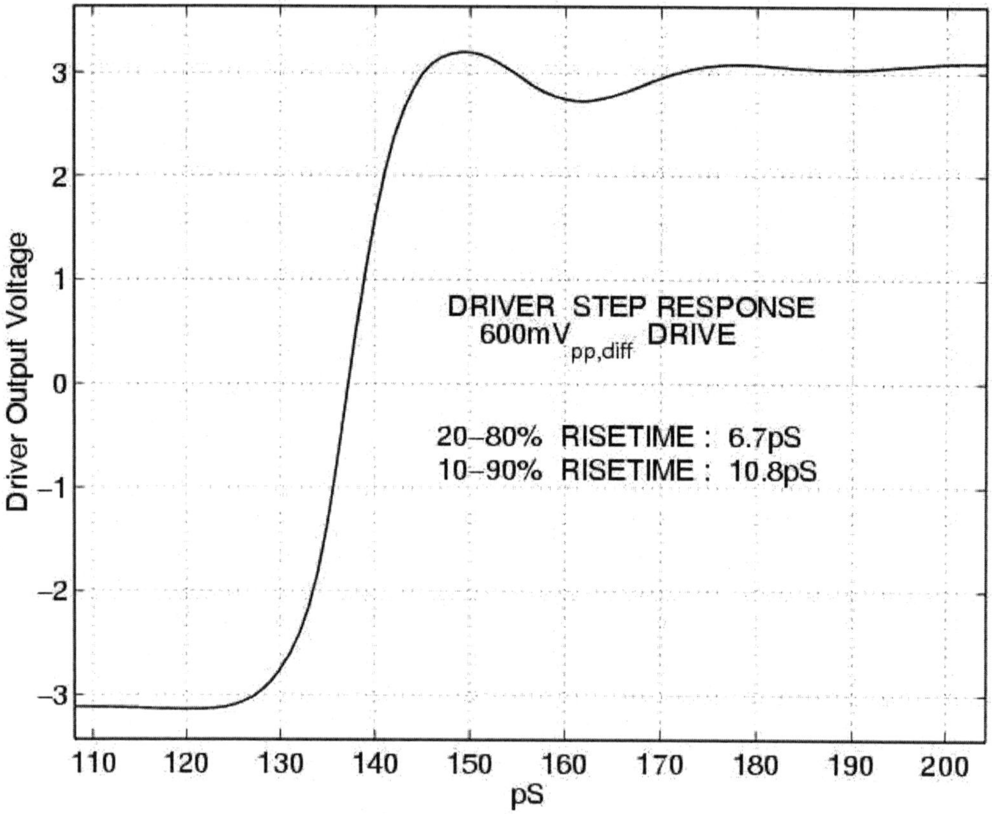

Fig. 17. Differential output step response of the driver.

References

1. J. P. Mattia, "An OC-768 transponder for 40 and 43 Gb/s short-reach links", *GaAs IC Symposium* (2003) 27–30.
2. M. Meghelli, A. Rylyakov, S. Zier, M. Sorna, D. Friedman and M. Beeson, " A 0.18 μm SiGe BiCMOS receiver and transmitter chipset for SONET OC-768 transmission systems", *IEEE Journal of Solid-State Circuits* **38** (2003) 2147– 2154.
3. H. Tao, D. Shaeffer, M. Xu, S. Benyamin, V. Condito, S. Kudszus, Q. Lee, A. Ong, A. Shahani, X. Si, W. Wong and M. Tarsia, "40-43-Gb/s OC-768 16:1 MUX/CMU chipset with SFI-5 compliance", *IEEE Journal of Solid-State Circuits* **38** (2003) 2169–2180.
4. W. Mouzannar, F. Jorge, S. Vuye, E. Dutisseuil and R. Lefevre, "40 Gbit/s high performances GaAs pHEMT high voltage modulator driver for long haul optical fiber communications", *GaAs IC Symposium* (2002) 163–166.
5. E. Legros,S. Vuye, L. Giraudet and C. Joly,"High-sensitivity 40 Gbit/s photoreceiver using GaAs PHEMT distributed amplifiers", *Electronics Letters* **34** (1998) 1351–1352.
6. D. Streit, R. Lai, A. Gutierrez-Aitken, M Siddiqui, B. Allen, A. Chau, W. Beale and A. Oki, " InP and GaAs components for 40 Gbps applications", *GaAs IC Symposium*

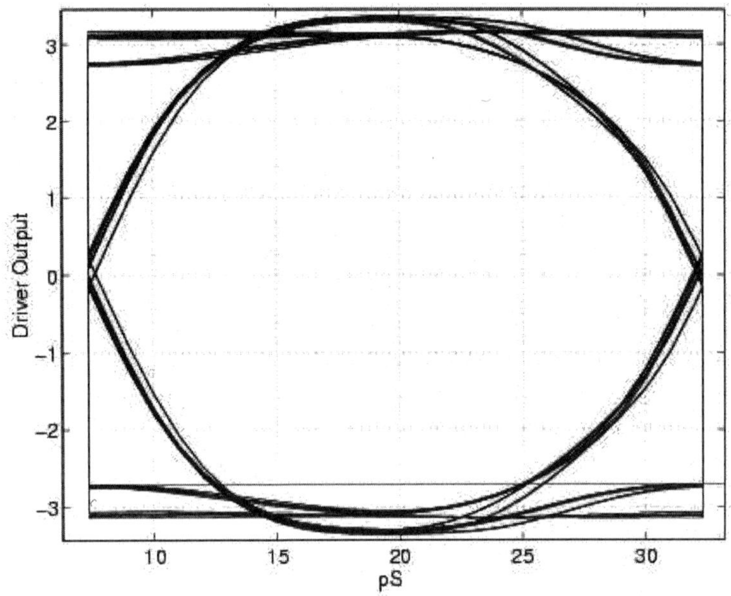

Fig. 18. Simulated eye diagram at the driver output at 40 Gbps.

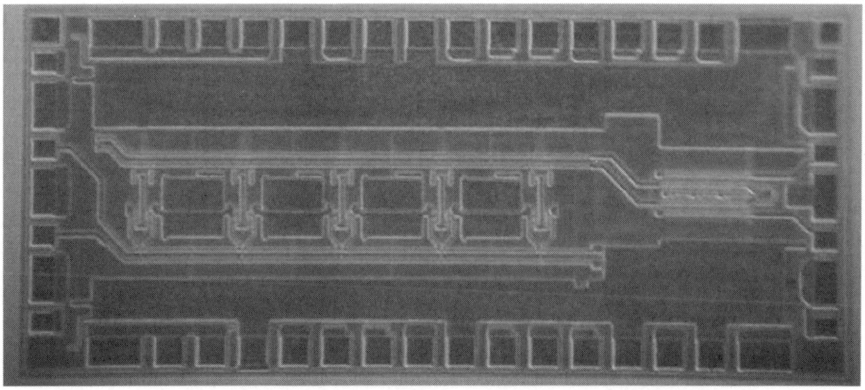

Fig. 19. Microphotograph of the chip.

(2001) 247–250.
7. H. Tran, F. Pera, D. S. McPherson, D. Viorel and S. P. Voinigescu, " 6-kΩ 43-Gb/s Differential Transimpedance-Limiting Amplifier With Auto-Zero Feedback and High Dynamic Range", *IEEE Journal of Solid-State Circuits* **39** (2004) 1680–1689.
8. R. Schmid, T. Meister, M. Rest and H. -M. Rein, "SiGe driver circuit with high output amplitude operating up to 23 Gb/s", *IEEE Journal of Solid-State Circuits* **34** (1999)

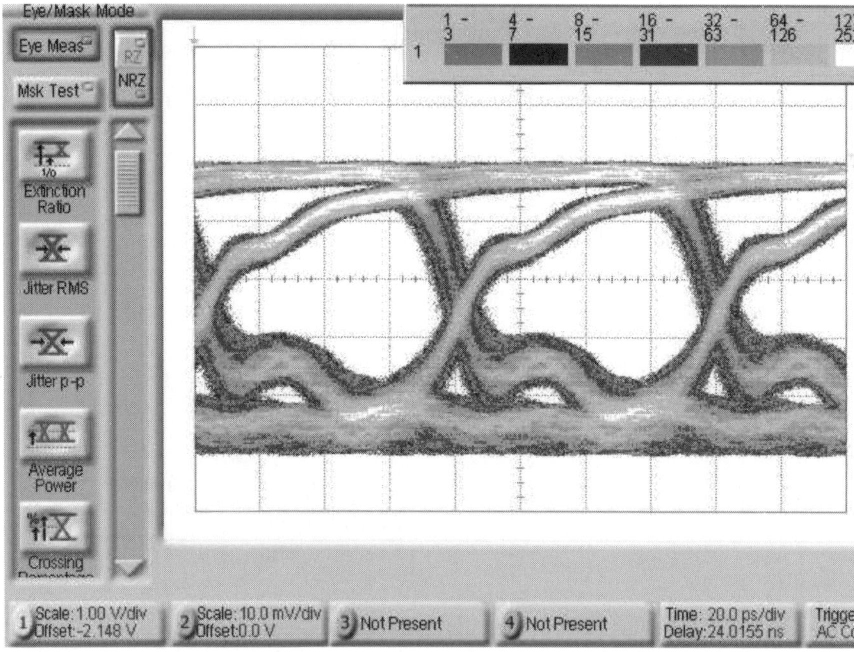

Fig. 20. Driver output eye-diagram for 12.5 Gbps data.

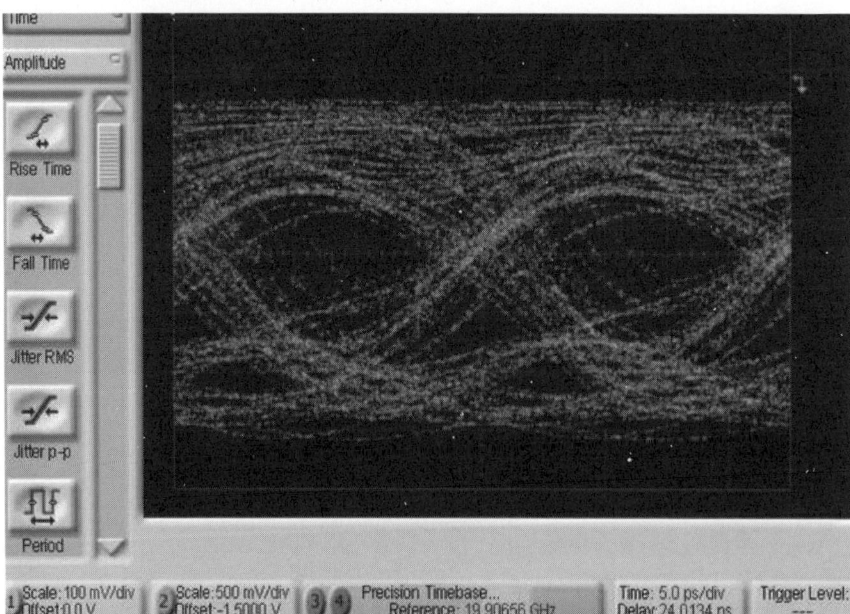

Fig. 21. Driver output eye-diagram for 40 Gbps data.

886–891.
9. M. Moller, T. Meister, R. Schmid, J. Rupeter, M. Rest, A. Schopflin and H. -M. Rein,"SiGe retiming high-gain power MUX for directly driving an EAM up to 50 Gbit/s", *Electronics Letters* **34** (1998) 1782–1784.
10. R. Schmid, T. Meister, M. Rest and H. -M. Rein,"40 Gbit/s EAM driver IC in SiGe bipolar technology", *Electronics Letters* **34** (1998) 1095–1097.
11. T. Wong, "Fundamentals of Distributed Amplification", *Artech House* (1992).

International Journal of High Speed Electronics and Systems
Vol. 15, No. 3 (2005) 497–512

Compact low-noise pulse generating lasers with repetition rates of 10 to 50 GHz

G. J. Spühler, L. Krainer, S. C. Zeller, Ch. Erny, R. Paschotta

Ultrafast Laser Physics, Institute of Quantum Electronic
Swiss Federal Institute of Technology
ETH Hönggerberg HPT, 8093 Zürich, Switzerland
spuehler@phys.ethz.ch

K. J. Weingarten

GigaTera
Technoparkstrasse 1, CH-8005 Zurich
Switzerland

U. Keller

Ultrafast Laser Physics, Institute of Quantum Electronic
Swiss Federal Institute of Technology
ETH Hönggerberg HPT, 8093 Zürich, Switzerland

We review the main results obtained with our recently introduced passively fundamental mode-locked Er:Yb:glass lasers mainly in the context of telecom applications, and we discuss their key enablers. Specifically, we focus on the aspects of the lasers for application in the time-domain for optical time-division multiplexing and in the wavelength domain for dense wavelength-division-multiplexing.

Keywords: Solid-state laser; passive mode locking; high pulse repetition rate.

1. Introduction

There are many compelling reasons to use return-to-zero (RZ) formats [1,2] and/or soliton dispersion management techniques [3,4] for novel transmission systems at 10 Gb/s, 40 Gb/s and higher bit rates per channel. These techniques rely on clean optical pulses and thus benefit greatly from the availability of simple, compact, low-noise and efficient optical pulse generators. Therefore, pulsed lasers are becoming increasingly important for telecom applications as data transmission rates continue to increase. There are many reasons to use a pulsed laser directly as a source in the transmitter of optical telecommunication systems, rather than an externally modulated cw source. Pulsed lasers eliminate the need for a high-end modulator to create the pulses and thereby simplify system architecture, increase efficiency, and reduce cost. Furthermore, the contrast ratio of pulsed lasers is typically much higher than for modulated cw sources. This improves system signal-to-noise and allows for further scaling to higher bit rates through optical time-division multiplexing (OTDM). Apart from the transmitter side, there are interesting applications of pulsed lasers also in the receivers of transmission systems, e.g. for demultiplexing and clock recovery [5-7]. In addition, pulse generating lasers (PGLs) can be used as multi-wavelength sources for dense wavelength-division-multiplexing (DWDM) systems [8,9]. Furthermore, there are applications outside the telecom area, in the fields of optical clocking [10,11], high-speed electro-optic sampling [12,13], frequency metrology (similar to the work presented in Ref. [14]) or generation of polarized electron beams for particle accelerators [15].

The following requirements have to be fulfilled for an ideal pulse generating source for telecom applications: Besides a repetition rate that is matched to the standards for 10, 40 Gb/s or higher transmission rates [16], reasonable average output power, low timing jitter, high contrast ratio and wavelength tunability in the regime of interest (mostly in the C-band from 1530 – 1565 nm) are required. The requirement for the pulse duration depends on the coding format. Usually the full width at half max (FWHM) of the pulses should be around a third of the bit period. Besides these requirements on the optical properties of the pulse train, high efficiency, compactness, simplicity, reliability and a cost-saving manufacturing process are important features for a telecom product.

In the past, pulse sources with pulse repetition rates of several GHz almost always involved either an edge-emitting semiconductor laser [17], which is usually actively or hybrid mode-locked, or a harmonically mode-locked fiber ring laser [18]. Edge-emitting semiconductor lasers appear very attractive due to their very compact and stable optical setup. However, the average optical output power is typically rather low and the timing jitter can be relatively high, degrading transmission at high bit rates. The increased timing jitter is explained by rather high cavity losses and a low average power [19]. Furthermore, these lasers often require a large amount of RF drive power for stable operation and are susceptible to satellite pulses generated at intracavity surfaces. Multi-GHz fiber lasers can easily generate high-quality pulses with good output power and low timing jitter. However, fiber lasers typically need to have a cavity length of several meters in order to get good pump absorption, gain, and thus efficient operation. Therefore sophisticated means are required to obtain stable harmonic mode locking with a large number of precisely equidistant pulses in the cavity. The individual pulses generated by harmonic mode locking do typically not exhibit a fixed phase relation. This excludes certain promising coding formats such as return-to-zero differential phase shift keying [20,21] which carry the data in the phase of the pulses, rather than in the amplitude. Additionally, also the typical fiber ring laser used for telecom applications makes use of an intracavity modulator which requires a large amount of RF power.

Here, we report on our recently developed alternative pulse sources for telecom applications, based on passively fundamental mode-locked solid-state lasers. Passive mode locking totally eliminates the use of a modulator and thus drastically reduces the demand on RF power. Only a rather weak RF signal (typically several orders of magnitude less than for driving a modulator) is required for feeding a phase-locked loop that keeps the cavity length of the passively mode-locked laser constant. Additionally, the typical diode-pumped passively mode-locked solid-state laser uses a rather low-loss cavity due to the limited small signal gain and the need to suppress Q-switched mode locking [22]. This leads intrinsically to low timing jitter, provided that a stable laser cavity setup is used. Fundamental mode locking in the multi-GHz regime leads to very compact cavity setups that can be implemented very ruggedly.

In the last millennium, the repetition rate of passively mode-locked solid-state lasers has been limited to a few GHz [23]. Especially in the telecom wavelength regime, where only few solid-state gain media are available cw-mode locking with multi-GHz pulse repetition rates have not been possible [24,25]. In recent years, the consequent exploitation of the flexibility of semiconductor saturable absorber mirrors (SESAMs) [26-28] in combination with increased understanding of the limitations [22,29-32] of cw mode locking allowed us to develop passively mode-locked lasers with multi-GHz pulse repetition rates, very good pulse quality, comparatively high output powers and wavelengths in the regime of interest. Passive mode locking implies that the pulses are achieved without using any multi-GHz electronics. Fundamental mode locking leads to ultra-compact cavity setups and to the fact that every output pulse is a copy of the same single pulse bouncing back and forth in the cavity. Therefore, pulse-to-pulse variations are minimized and the output pulses have a fixed phase relation.

In this paper we limit our discussion to the 1.5 μm wavelength regime. However, much progress has also been achieved at other wavelengths, important for different applications as for example optical clocking. A recent review paper gives more information with respect to that [33].

2. Er:Yb:glass

The choice of solid-state gain media in the wavelength regime of interest is rather limited. Mainly Er:Yb:glass and Cr:YAG are used. The latter has very strict requirements on the pump source due to its weak pump absorption and it exhibits relatively low efficiencies, because of limited crystal quality and relatively high induced losses [34,35]. Therefore, the so far most successful gain medium for a multi-GHz passively mode-locked solid state laser in the telecom wavelength regime is Er:Yb:glass, although it has very small cross sections and thus a strong tendency towards Q-switched mode locking (QML) [22]. The way we overcome these limitations is explained in Section 3. Generally, Er:Yb:glass is well suited for telecom applications. Its gain bandwidth covers essentially the whole C-band [36] and allows for ultrashort pulse-generation [37], it can be efficiently pumped with standard telecom-grade 980-nm laser diodes also used in erbium-doped fiber amplifiers (EDFAs), it can be produced in excellent quality and large amounts, and it is resistant and cheap. An alternative way to reach the 1.5-μm regime is to use nonlinear processes. Synchronously pumped optical parametrical oscillators (OPO) [38,39] are powerful tools, which give higher power levels and wider tuning ranges at the cost of increased complexity, reduced stability and efficiency.

3. Challenges – Q-switched mode locking (QML)

Passively mode-locked solid-state lasers typically have a pulse repetition rate of a few hundred MHz. For fundamental mode locking, the inverse pulse repetition rate is given by the time a pulse needs for one round-trip in the laser cavity. The shorter the resonator the higher the repetition rate - and (for constant average output power) the smaller the intracavity pulse energy. As mentioned, the main challenge one has to face when pushing the repetition rate of a passively mode-locked laser to the multi-GHz regime, is the suppression of QML. QML is a state of operation where the relaxation oscillations are destabilized; the damping effect for increased intracavity powers due to gain saturation is not strong enough to suppress the growth due to increased saturation of the absorber. A high repetition rate naturally leads to relatively low intracavity pulse energies which limit the saturation of the gain and thus increases the tendency for QML. In addition, Er:Yb:glass exhibits rather small cross sections and has thus a high saturation energy which again increases the tendency towards QML.

In order to suppress QML in our Er:Yb:glass lasers we rely on the following concepts: The SESAM allows to custom design the main absorber parameters within a certain range, mainly by adopting the structure of the device and its composition [27,40]. We exploit that range for minimized saturation fluence and modulation depth which both helps to suppress QML [22]. Furthermore, we make use of a roll-over in the saturation curve of our SESAMs. The reflectivity of a semiconductor SESAM is generally expected to increase with increasing pulse energy. However, for higher pulse energies the reflectivity can decrease again and exhibit a roll-over in the nonlinear reflectivity curve caused by inverse saturable absorption [31]. This effectively increases the losses for pulse fluences above this roll-over point and acts as a pulse energy limiter and thus helps to suppress QML [30].

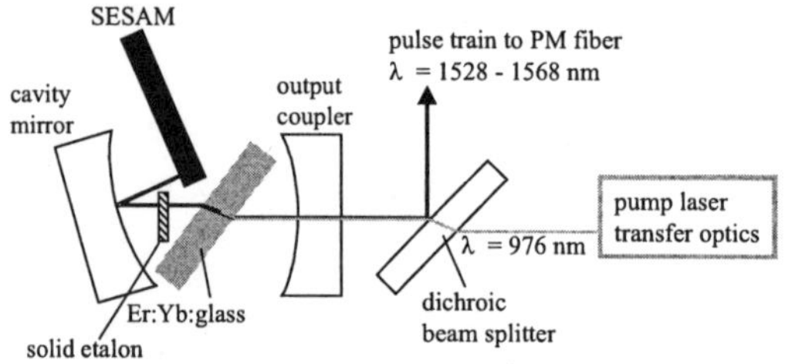

Fig. 1 Setup of the 10-GHz Er:Yb:glass laser. Cavity length ≈ 14 mm.
50-GHz laser has similar setup with cavity length ≈ 3 mm and without etalon.

For fundamental mode locking, the repetition rate dictated by the application determines the cavity length. With the constraint of the length of the cavity and the SESAM being a flat end mirror of the laser, we designed the cavity in such a way that its mode size on the SESAM in combination with intracavity pulse energy leads to a pulse fluence on the absorber which is close to the roll-over point. The mode size in the gain medium is minimized taking into account the pump beam quality and the thickness of the gain element, leading to a simple three-mirror-cavity as shown in Fig. 1. This cavity setup minimizes the saturation energy of the gain and maximizes the efficiency and the exploitation of the inverse saturable absorption insuring the suppression of QML. The laser glass is inserted into the cavity under Brewster's angle to induce linear polarization and minimize intracvity losses. A further important point is the use of a single-mode 980-nm pump diode. It enables optimum mode matching of pump and laser mode even though the laser mode size in the gain is very small. This allows for good efficiency and excellent mode quality.

With this approach of diode-pumped passively fundamental mode-locked Er:Yb:glass lasers, we have presented pulse repetition rates up to 50 GHz. Fig. 1 shows the cavity design. For a 10-GHz laser the total cavity length is 14 mm, and for the 50-GHz cavity it is as short as 3 mm.

4. Results

4.1. Time domain

In time domain applications, the user is mainly interested in the temporal evolution of the output of the laser: Pulse repetition rate, pulse duration, extinction ratio, timing jitter, etc. Each pulse is used for a special purpose: to carry information [41] or to induce some switching window [7] or for other processes. The center wavelength and width of the optical spectrum as well as its purity are of interest too, especially for systems with multiple channels.

4.1.1. 10-GHz pulse generating lasers

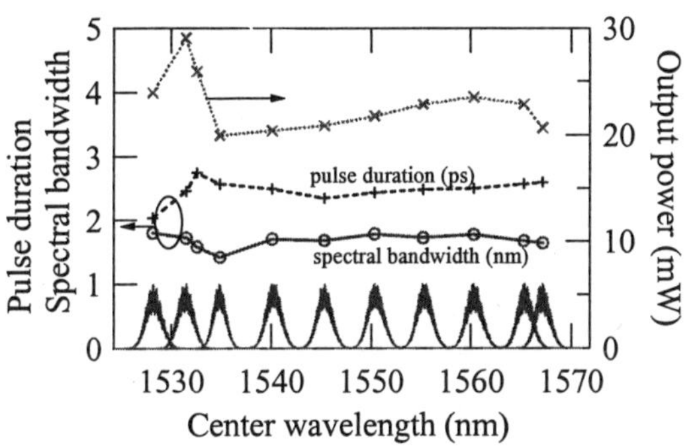

Fig. 2 Pulse duration (dashed), spectral bandwidth (solid) and output power (dashed-dotted) as a function of center wavelength for a 10-GHz Er:Yb:glass laser together with several sample spectra.

Fig. 1 shows the cavity set up for a 10-GHz PGL, as explained in section 3. Next to the SESAM and the Er:Yb:glass gain element, the cavity contains a solid etalon to select the center wavelength. Varying the SESAM parameters and the intracavity filter effect from the etalon, pulse durations from 1 to 19 ps could be obtained [42,43]. With the etalon the laser wavelength can be adjusted within 1528 – 1568 nm (full C-band) by tuning the tilt angle (Fig. 2). During tuning, only week variations of the main laser parameters, such as pulse duration, spectral bandwidth and average power are observed [43]. We typically couple the output of the laser into a polarization-maintaining fiber where we get 10 to 20 mW of average power with a polarization extinction ratio of >20 dB for a pump power of about 250 mW.

The output pulse train is charcterized by a very high contrast ratio of typically >30 dB and a very stable optical spectrum (Fig. 3). Fig. 3a shows a high dynamic range autocorrelation with a scan range exceeding the pulse separation. This reveals the first cross correlation. The highest signal between the pulses is 36 dB below the carrier showing the excellent contrast ratio that can be achieved with these lasers. The measurement is shown together with a Gaussian and a $sech^2$ fit, indicating that the measured pulses are fitted better by a Gaussian pulse. In order to show the suitability of the Er:Yb:glass laser for transmission at high bit rates, we show in Fig. 4 an autocorrelation of a pulse train which has been obtained by optical time division multiplexing (OTDM) of a 10-GHz laser up to 160 GHz with commercially available multiplexers (PriTel Inc., OCM-4). First of all, we can see that the pulses are short enough (in this case ≈1 ps) for such a high bit rate. Secondly, the contrast ratio is high enough even for 160 Gb/s transmission to avoid a significant noise level between the multiplexed pulses.

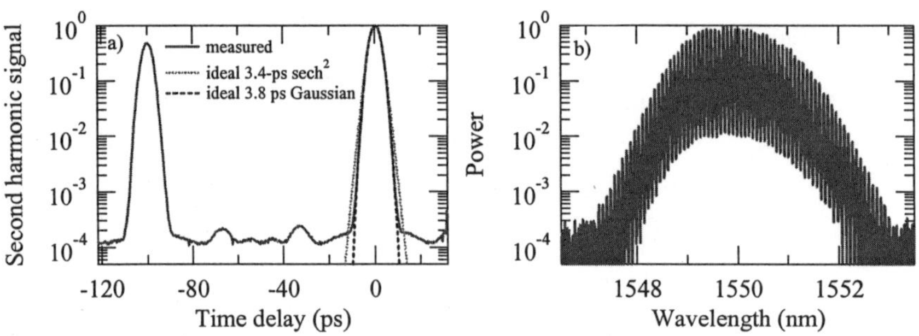

Fig. 3 a) Slow scan autocorrelation. The autocorrelation scan range exceeds the pulse separation. Maximum signal level between the pulses is –36 dBc. The Gaussian fits the measured autocorrelation trace better than the autocorrelation of a $sech^2$ pulse. b) Typical optical spectrum of a 10-GHz Er:Yb:glass laser taken with a resolution of 0.01 nm resolving the individual longitudinal modes.

Of further importance for transmission systems with such high bit rates is synchronization of the pulses of the PGL to an existing clock signal with very low phase noise. This synchronization is naturally given for actively or hybrid mode-locked lasers, at the price of the need for an intracavity high speed modulator, which requires a lot of drive power. A passively mode-locked laser does not need any active mean to generate the pulses, but to synchronize them to the clock signal. This can be done by a simple phase-locked-loop (PLL), which locks the cavity length to the clock signal [44,45]. This approach of passive mode locking – active synchronization has several advantages: The amount of power required from the clock is orders of magnitude smaller than required for a modulator and the simplicity of the pulsing mechanism allows the clock signal to drift within a certain range while the laser remains phase locked. This flexibility is beneficial for applications in the receiver end of a transmission system, as during transmission the clock signal can be deteriorated significantly [7]. For synchronization to an external clock signal, we mount the SESAM on a piezo element. A small part of the output signal is tapped and its phase is compared to the one of the clock signal. The error signal is amplified, filtered and fed back to the piezo element to adjust the cavity length for minimum phase deviation. With a relatively low loop bandwidth of less than 20 kHz, we regularly obtain rms timing jitter values below 80 fs, when integrating the phase noise power spectral density measured with the von der Linde technique [46] from 10 Hz to 1 MHz (Fig. 5). The measurement is mainly limited by the noise of the local oscillator of the microwave spectrum analyzer rather than the laser itself. The measured phase noise is by far low enough to perform transmission experiments at 160 Gb/s [41]. As we apply fundamental mode locking, there are no supermodes.

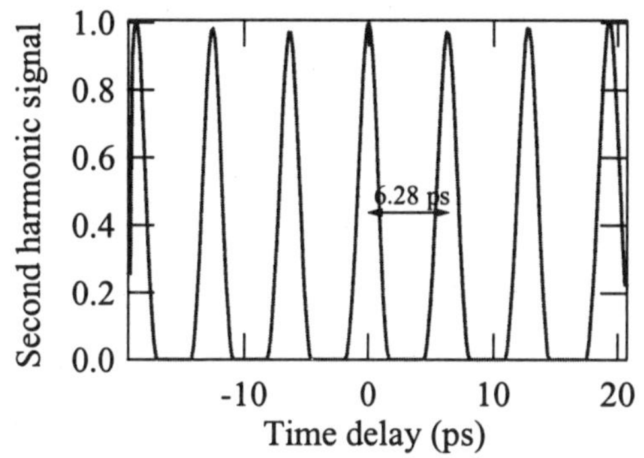

Fig. 4 Autocorrelation trace of a pulse train multiplexed from 10 GHz to 160 GHz.

Recently, the theory of noise of mode-locked lasers has been investigated in detail by Paschotta [19,47]. In an ideal case, the mechanical and electrical implementation of the laser would exclude any mechanical vibrations of the optics and pump power fluctuations. Then, laser noise is limited by the effect of quantum noise. In simple cases, the timing noise spectral power density of mode-locked lasers as an effect of quantum noise scales with small signal gain, the square of the pulse duration and inversely with the intracavity pulse energy and round-trip time [19]. This explains why mode-locked solid-state lasers have a great potential to exhibit low phase noise. They usually operate with low cavity losses and thus have a small saturated gain value. Additionally the rather high obtainable power levels lead to a relatively large intracavity pulse energy. For edge-emitting external cavity semiconductor lasers the situation is typically different: They suffer from high cavity losses and thus have a high saturated gain. Additionally, they have strong power limitations keeping the intracavity pulse energy low. Thus their quantum noise induced phase noise is relatively high compared to mode-locked solid-state lasers. On the other hand, the cavity setup is typically more rugged so that mirror vibrations play less of a role. The third candidate for PGLs with high repetition rate, the mode-locked fiber ring laser profits from its long cavity length, leading to a long round-trip time. This can more than compensate for the relatively high cavity losses. However, special care has to be taken to suppress supermode noise due to harmonic mode locking 48.

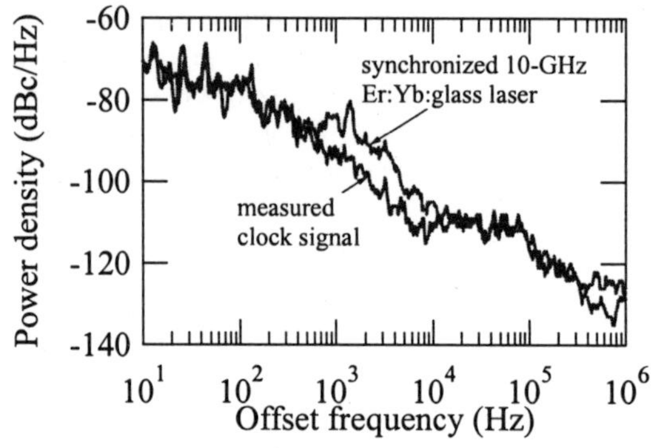

Fig. 5 Phase noise power spectral density for clock signal and synchronized Er:Yb:glass laser measured with von der Linde method. Apart from a narrow window around 1 kHz the laser noise equals the measured clock noise which is mainly limited by the local oscillator in the microwave spectrum analyzer. The integrated (10 Hz – 1 MHz) phase noise yields an rms timing jitter of below 80 fs.

The repetition rate of the PGL's pulses for a 10-Gb/s connection is defined by the applied standard. For example, SONET (Synchronous Optical NET) STS-192 requires a repetition rate of 9.95328 GHz. However, in practice there are a number of different data rates possible. If forward error correction (FEC) is applied, the required pulse repetition rate has to be adjusted to 10.667 GHz, or even higher with more sophisticated FEC coding. A pulse source with continuously tunable repetition rate increases the flexibility of a network and allows for later upgrades to different transmission standards. The complex setup of a harmonically mode-locked fiber ring laser requires readjustment of several other parameters when adjusting the repetition rate. This is significantly simpler for hybrid mode-locked semiconductor lasers.

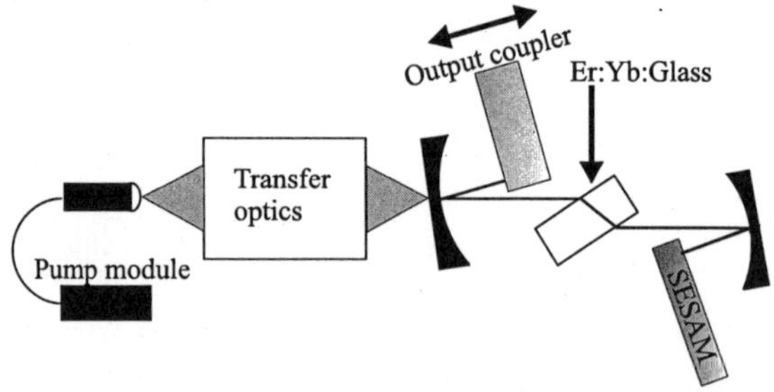

Fig. 6 Setup of the repetition rate tunable four mirror 10-GHz Er:Yb:glass laser.

We refined the three-mirror-cavity setup (Fig. 1) by using a four-mirror-design (Fig. 6). By the addition of a flat mirror the cavity can be arranged in such a way that a large laser mode is obtained on the flat output coupler. This allows to tune the repetition rate simply by moving this end mirror along the beam axis. Due to the large laser mode on the flat output coupler, the beam divergence is very small. Thus the stability region with respect to the position of the output coupler is very wide, and movement of it only weakly affects the mode sizes in the gain medium and on the saturable absorber. Therefore, the main laser parameters (pulse duration, output power and optical bandwidth) hardly change, and the tuning range is mainly limited by geometric factors (i.e. achievable travel range). This setup significantly reduces the dependence of the laser mode size in the gain, on the SESAM and in the output beam on repetition rate, thus avoiding problems like changing output power, beam quality, pulse duration, or fiber coupling efficiency. With this cavity setup, we can vary the pulse repetition rate between 8.8 GHz to 13.3 GHz with negligible variation of the main pulse parameters (Fig. 7) [49]. This tuning range allows the laser to be used for all the different RZ transmission standards at 10 Gb/s as well as a multi-wavelength source for the ITU channel spacing of 12.5 GHz (see Section 4.2).

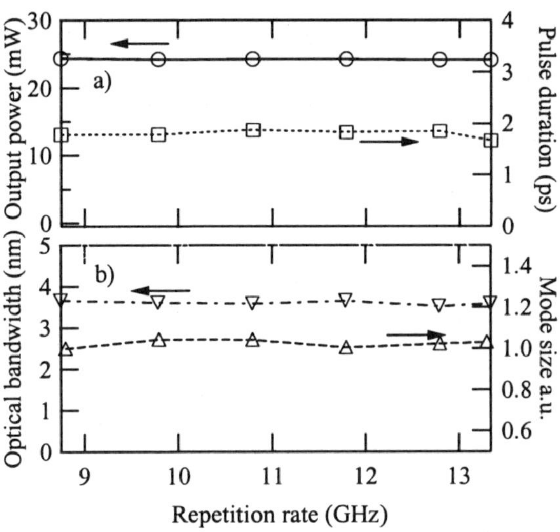

Fig. 7 a) Output power (solid) and pulse duration (dotted) and b) optical bandwidth (dashed-dotted) and relative output beam size (dashed) as a function of repetition rate. The main pulse parameters remain essentially unchanged during tuning the repetition rate from 8.8 to 13.3 GHz.

4.1.2. 40-GHz pulse generating lasers

We have also developed an Er:Yb:glass laser directly generating a pulse train with 40-GHz repetition rate [50]. The laser is based on the same principles and the same cavity setup as shown in Fig. 1. Also the pulse parameters are similar to the 10-GHz laser (18-mW average output power, 4.3-ps pulses). Due to space constraints in the extremely short cavity, we did not prove wavelength or pulse repetition rate tunability yet. For space reasons, for repetition rates higher than 10 GHz, we used 40% thinner Er:Yb:glass Brewster gain elements. Due to the somewhat decreased pump absorption and the inferior mirror quality of the more strongly curved mirrors, the overall efficiency decreased slightly at these repetition rates. The reduced interaction length of pump and gain could theoretically be compensated for by the use of adjusted Yb and Er concentrations of the gain element.

4.2. Wavelength domain

In the wavelength domain, fundamental-mode-locked lasers generate a stable comb-shaped optical spectrum (Fig. 3b), where the spacing of the individual longitudinal modes exactly equals the pulse repetition rate and shows a high optical signal-to-noise ratio (OSNR). Therefore, pulse-generating lasers are intrinsically wavelength-stable multi-channel sources which can be used as comb-generating lasers in dense wavelength-division-multiplexing (DWDM) [8,9]. Varying the cavity length and comparing/locking a single longitudinal mode of the laser to an ITU grid line (or any other reference), the entire optical comb is automatically locked to the ITU grid. The distinct channels of the laser can be separated with appropriate means (e.g. an arrayed wave guide grating [51]), individually modulated and recombined to launch into the network [52]. Compared to the classic DWDM approach with one fixed-wavelength-laser per channel, the comb-generating laser approach brings many benefits to the system, such as decreased number of components (i.e. only one set of driver electronics and one wavelength locker are required), smaller inventory, space and power requirements, etc.

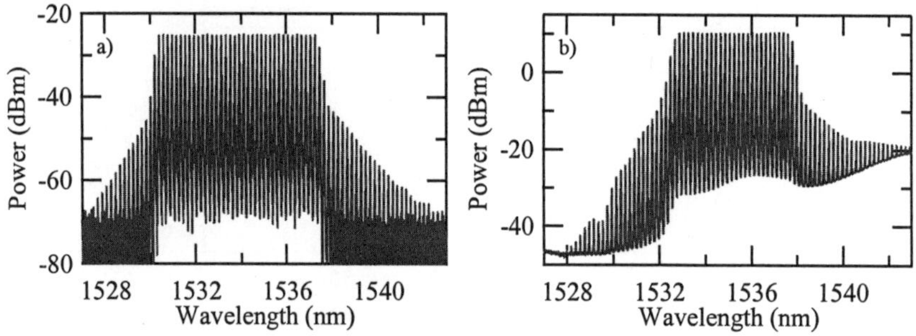

Fig.8 Optical spectrum from 25-GHz a) laser flattened with DGE and b) flattened with DGE and amplified with an EDFA.

As current channel spacing standards are 25 GHz, 50 GHz, 100 GHz, or higher (International Telecommunication Union, ITU), we developed a 25 GHz and a 50 GHz version of the Er:Yb:glass laser [53,54]. Fig. 1 still holds for the cavity designs. The overall cavity length is then about 6 mm and 3 mm respectively. Also the output parameters are very similar to the 10-GHz and 40-GHz lasers: several mW average power, about 2-ps pulse duration, 30-dB contrast ratio, and a 2-3 nm of spectral FWHM. We could tune the wavelength of the 25-GHz laser in the range of 1528 – 1561 nm. Tilting the etalon moves the envelope of the spectrum across the C-band, whereas changing the cavity length moves the individual longitudinal modes underneath the envelope. In this way, the comb spectrum can be tuned to any desired wavelength in the C-band, i.e. to the ITU grid, essentially without changing the grid spacing. With a 25-GHz channel spacing, the maximum frequency offset between a longitudinal mode and a required reference line is

12.5 GHz. To remove this offset, the cavity length has to be changed by a quarter wavelength leading to a change in repetition rate (and thus channel spacing) of only 1.6 MHz (four times more for the 50-GHz laser), which is well below the tolerance of the ITU channels and typical bandwidths of DFB lasers. The reason for this small change in repetition rate is that the carrier frequency is several orders of magnitude higher than the repetition rate itself. Therefore, a sub-wavelength change in cavity length has only a minimum impact on the channel spacing.

For a DWDM application each channel (longitudinal mode) should have the same power rather than the Gaussian like spectrum of a typical mode-locked laser. Therefore, we combined the 25-GHz laser with a dynamic gain equalizer (DGE) (Silicon Light Machines, Model:SLM2200). The DGE allows setting the attenuation of each channel individually and in real time, controlled by a computer. Fig. 8a shows the optical spectrum of the laser (without etalon) in conjunction with the DGE. A fraction of the output is detected and checked for flatness. This gives an error signal which is fed back to the DGE. Using the maximum dynamic range of the DGE (17 dB), we could get up 32 channels with a flatness of better than 0.4 dB. In order to increase the power per channel, we added an EDFA to the system (after the DGE). The gain narrowing reduced the number of channels to 25, however each carrying 10 dBm average power, and an OSNR exceeding 35 dB (Fig. 8b). No additional filters have been used to suppress the amplified spontaneous emission of the EDFA. The system is very stable, as the original laser spectrum is stable and adaptive flattening is applied in addition.

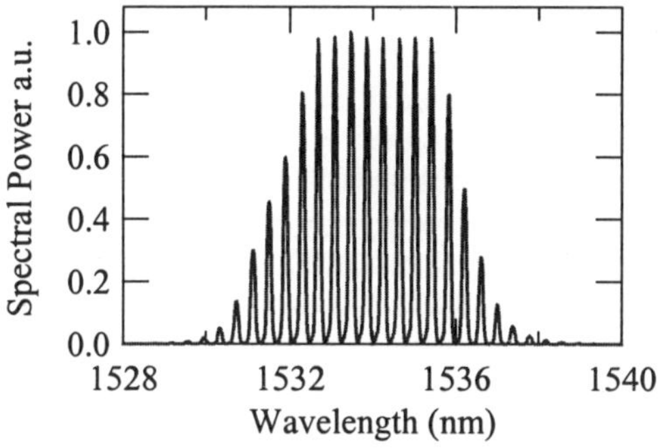

Fig.9 Optical spectrum from 50-GHz laser flattened with DGE.

Increasing the pulse repetition rate leads to a wider channel spacing. This relaxes the demands on the filter characteristics of the channel add/drop nodes, and it allows for higher channel capacities. Therefore, we have developed a 50-GHz Er:Yb:glass laser [54]. For locking the longitudinal modes of this 50-GHz comb to the ITU grid, we used a simple feedback loop consisting of a wavelength locker from JDS Uniphase, a simple integrating circuit, and a piezo below the SESAM to tune the cavity length. Fig. 9 shows the optical spectrum of the multi wavelength source, locked to the ITU grid. Here, for flattening we used a somewhat simpler DGE (ZettaManager 12-band from ZettaLight) with less maximum dynamic range (10 dB) and a coarser channel spacing. Therefore, only up to 10 channels with a flatness <1 dB have been generated. The power in each of the equalized modes is –25 dBm. The measured optical signal-to-noise ratio (OSNR) was 65 dB, probably limited by the spectrometer.

Keeping in mind the simplicity of this approach, its potential for DWDM system applications becomes clear. An additional stage for super-continuum generation can even increase the number of obtained channels substantially [52]. The super-continuum generation would benefit from the spectral purity and stability of the presented laser. In first super-continuum generation experiments carried out at 10 GHz we obtained a stable smooth spectrum ranging from 1450 nm to 1650 nm. For that the pulse train from the laser was amplified in an EDFA and launched into a highly nonlinear fiber, generating a huge number of additional longitudinal modes, still exactly spaced by the repetition rate of the seed laser.

5. Conclusion

We have presented various novel multi-GHz sources, all based on diode-pumped, passively fundamental mode-locked Er:Yb:glass lasers. The application of a semiconductor saturable absorber mirror in combination with optimized cavity designs allowed progression in performance regimes where these lasers enable and/or improve new approaches for telecom applications. The combination of passive mode locking of a solid-state laser with active synchronization leads to outstanding pulse properties in a simple and cheap setup. Main characteristics are high contrast ratio, short pulses, clean spectrum with high signal to noise ratio, and low timing jitter. The main pulse parameters can be varied in a wide range. I.e. turn key tuning of the wavelength throughout the C-band and the repetition rate covering all the different RZ transmission standards at 10 Gb/s has been presented. Pulse durations between 1 ps and 19 ps and pulse repetition rates of 10, 25, 40, and 50 GHz enable OTDM and DWDM applications.

For future development, repetition rates beyond 80 and 100 GHz seem to be appealing for higher bit rates or larger channel spacings. Possible limiting factors are the extremely small cavity dimensions and Q-switching instabilities. Also introduction of a tuning element in the 40 and 50-GHz lasers is a challenge. Further, we believe that with improved mechanical design and phase-locked-loop electronics, the rms timing titter can still be significantly reduced.

Acknowledgments

The authors would like to acknowledge financial support through Hasler Stiftung and the Kommission für Technologie und Innovation (KTI).

References

1 H. Taga, M. Suzuki, and Y. Namihira, "Polarization mode dispersion tolerance of 10Gbit/s NRZ and RZ optical signals", *Electron. Lett.* **34** (1998) 2098-100.

2 L.-P. Chen and C.-X. Shi, "Experimental Comparison of Transmission System Penalty of 10 Gbit/s RZ and NRZ Optical Signals Due to Polarization-Mode Dispersion", *Microwave and Optical Technology Letters* **25** (2000) 297-300.

3 L. F. Mollenauer and P. V. Mamyshev, "Massive wavelength-division multiplexing with solitons", *IEEE J. Quantum Electron.* **34** (1998) 2089-2102.

4 M. Nakazawa, H. Kubota, K. Suzuki, E. Yamada, and A. Sahara, "Ultrahigh-Speed Long-Distance TDM and WDM Soliton Transmission Technologies", *IEEE J. Sel. Topics in Quantum Electron.* **6** (2000) 363-396.

5 S. Bigo, O. Leclerc, and E. Desurvire, "All-Optical Fiber Signal Processing and Regenration for Soliton Communications", *IEEE J. Sel. Topics in Quantum Electron.* **3** (1997) 1208-23.

6 T. Yamamoto, L. K. Oxenløwe, C. Schmitt, C. Schubert, E. Hilliger, U. Feiste, J. Berger, R. Ludwig, and H. G. Weber, "Clock recovery from 160 Gbit/s data signals using phase-locked loop with interferometric optical switch based on semiconductor optical amplifier", *Electron. Lett.* **37** (2001) 509-10.

7 L. Oxenløwe, A. Siahlo, K. Berg, A. T. Clausen, B. Sørensen, K. Yvind, P. Jeppesen, K. Hansen, K. Hoppe, and J. Hanberg, "A novel 160 Gb/s receiver configuration including a glass crystal pulsed laser, photonic crystal fiber and a simple dynamic clock", in *29th European Conference on Optical Communication*, Vol. 6 (B. Costa, C. Someda, Rimini, Italy, 2003).

8 E. A. D. Souza, M. C. Nuss, W. H. Knox, and D. A. B. Miller, "Wavelength-division multiplexing with femtosecond pulses", *Opt. Lett.* **20** (1995) 1166-1168.

9 L. Boivin, M. Wegmueller, M. C. Nuss, and W. H. Knox, "110 Channels x 2.35 Gb/s from a single femtosecond laser", *IEEE Photon. Technol. Lett.* **11** (1999) 466-468.

10 D. A. B. Miller, "Optics for low-energy communication inside digital processors: quantum detectors, sources, and modulators as efficient impedance converters", *Opt. Lett.* **14** (1989) 146-148.

11 A. Bhatnagar, C. Debaes, R. Chen, N. C. Helman, G. A. Keeler, D. Agarwal, H. Thienpont, and D. A. B. Miller, "Receiver-less clocking of a CMOS digital circuit using short optical pulses", in *2002 IEEE/LEOS Annual Meeting; Conference Proceedings, 15th Annual Meeting of the IEEE Lasers & Electro-Optics Society*, Vol. 1, edited by F. X. Kärtner (Glasgow, Scotland, 2002), pp. 127-8.

12 J. A. Valdmanis and G. A. Mourou, "Subpicosecond electrooptic sampling: principles and applications", *IEEE J. Quantum Electron.* **22** (1986) 69-78.

13 K. J. Weingarten, M. J. W. Rodwell, and D. M. Bloom, "Picosecond optical sampling of GaAs integrated circuits", *IEEE J. Quantum Electron.* **24** (1988) 198-220.

14 H. S. Margolis, C. S. Edwards, G. P. Barwood, and P. Gill, "A passive optical frequency comb for frequency metrology in the 1.5 micron optical communications band", in *2002 IEEE/LEOS Annual Meeting; Conference Proceedings, 15th Annual Meeting of the IEEE Lasers & Electro-Optics Society*, Vol. 1, edited by F. X. Kärtner (Glasgow, Scotland, 2002), pp. 229-30.

15 A. Hatziefremidis, D. N. Papadopoulos, D. Fraser, and H. Avramopoulos, "Laser sources for polarized electron beams in cw and pulsed accelerators", *Nucl. Instrum. Meth. A* **431** (1999) 46-52.

16 TechFest, (http://www.techfest.com/networking/wan/sonet.htm, 2003).

17 K. Sato, "Semiconductor light sources for 40-Gb/s transmission systems", *IEEE J. Lightwave Technol.* **20** (2002) 2035-2043.

18 E. Yoshida, N. Shimizu, and M. Nakazawa, "A 40-GHz 0.9-ps regeneratively mode-locked fiber laser with a tuning range of 1530-1560 nm", *IEEE Photon. Technol. Lett.* **11** (1999) 1587-1589.

19 R. Paschotta, "Noise of mode-locked lasers. Part II: Timing jitter and other fluctuations", *Appl. Phys. B* **79** (2004) 163-173.

20 J. Leibrich, C. Wree, and W. Rosenkranz, "CF-RZ-DPSK for Suppression of XPM on Dispersion-Managed Long-Haul Optical WDM Transmission on Standard Single-Mode Fiber", *IEEE Photon. Tech. Lett.* **14** (2002) 155-7.

21 C. Xu, X. Liu, L. F. Mollenauer, and X. Wei, "Comparison of Return-to-Zero Differential Phase-Shift Keying and On-Off Keying in Long-Haul Dispersion Managed Transmission", *IEEE Photon. Tech. Lett.* **15** (2003) 617-9.

22 C. Hönninger, R. Paschotta, F. Morier-Genoud, M. Moser, and U. Keller, "Q-switching stability limits of continuous-wave passive mode locking", *J. Opt. Soc. Am. B* **16** (1999) 46-56.

23 U. Keller, "Ultrafast solid-state lasers", *Progress in Optics* **46** (2004) .

24 B. C. Collings, K. Bergman, and W. H. Knox, "True fundamental solitons in a passively mode-locked short cavity Cr^{4+}:YAG laser", *Opt. Lett.* **22** (1997) 1098-1100.

25 P. Laporta, S. Longhi, M. Marchesi, S. Taccheo, and O. Svelto, "2.5 GHz and 5 GHz Harmonic Mode-Locking of a Diode-Pumped Bulk Erbium-Ytterbium Glass Laser at 1.5 Microns", *IEEE Photon. Technol. Lett.* **7** (1995) 155-157.

26 U. Keller, D. A. B. Miller, G. D. Boyd, T. H. Chiu, J. F. Ferguson, and M. T. Asom, "Solid-state low-loss intracavity saturable absorber for Nd:YLF lasers: an antiresonant semiconductor Fabry-Perot saturable absorber", *Opt. Lett.* **17** (1992) 505-507.

27 U. Keller, K. J. Weingarten, F. X. Kärtner, D. Kopf, B. Braun, I. D. Jung, R. Fluck, C. Hönninger, N. Matuschek, and J. Aus der Au, "Semiconductor saturable absorber mirrors (SESAMs) for femtosecond to nanosecond pulse generation in solid-state lasers", *IEEE J. Sel. Top. Quantum Electron.* **2** (1996) 435-453.

28 U. Keller, "Recent developments in compact ultrafast lasers", *Nature* **424** (2003) 831-838.

29 R. Paschotta and U. Keller, "Passive mode locking with slow saturable absorbers", *Appl. Phys. B* **73** (2001) 653-662.

30 A. Schlatter, S. C. Zeller, R. Grange, R. Paschotta, and U. Keller, "Pulse energy dynamics of passively mode-locked solid-state lasers above the Q-switching threshold", *J. Opt. Soc. Am. B* **21** (2004) 1469-1478.

31 R. Grange, M. Haiml, R. Paschotta, G. J. Spühler, L. Krainer, M. Golling, O. Ostinelli, and U. Keller, "New regime of inverse saturable absorption for self-stabilizing passively mode-locked lasers", *Appl. Phys. B* in press (2004) .

32 M. Haiml, R. Grange, and U. Keller, "Optical characterization of semiconductor saturable absorbers", *Appl. Phys. B* **79** (2004) 331-339.

33 R. Paschotta, L. Krainer, S. Lecomte, G. J. Spühler, S. C. Zeller, A. Aschwanden, D. Lorenser, H. J. Unold, K. J. Weingarten, and U. Keller, "Picosecond pulse sources with multi-GHz repetition rates and high output power", *to appear in New Journal of Physics* (2004) .

34 A. Sennaroglu, "Analysis and optimization of lifetime thermal loading in continuous-wave Cr^{4+}-doped solid-state lasers", *J. Opt. Soc. Am B* **18** (2001) 1578-86.

35 A. Sennaroglu, "Optimization of power performance in room-temperature contunuous-wave Cr^{4+}:YAG lasers", *Opt. Comm.* **192** (2001) 83-9.

36 R. Francini, F. Giovenale, U. M. Grassano, P. Laporta, and S. Taccheo, "Spectroscopy of Er and Er±Yb-doped phosphate glasses", *Optical Materials* **13** (2000) 417-425.

37 G. J. Spühler, L. Krainer, E. Innerhofer, R. Paschotta, K. J. Weingarten, and U. Keller, "Soliton mode-locked Er:Yb:glass laser", *Opt. Lett.* in press (2004).

38 S. Lecomte, R. Paschotta, M. Golling, D. Ebling, and U. Keller, "Synchronously pumped optical parametric oscillators in the 1.5-μm spectral region with a repetition rate of 10 GHz", *J. Opt. Soc. Am. B* **21** (2004) 844-850.

39 S. Lecomte, R. Paschotta, S. Pawlik, B. Schmidt, K. Furusawa, A. Malinowski, D. J. Richardson, and U. Keller, "Synchronously pumped optical parametric oscillator with a repetition rate of 81.8 GHz", *submitted to Photonics Technology Letters* (2004).

40 U. Keller, in *Nonlinear Optics in Semiconductors*, Vol. 59, edited by E. Garmire and A. Kost (Academic Press, Inc., Boston, 1999), pp. 211-286.

41 J. P. Turkiewicz, E. Tangdiongga, G. D. Khoe, H. d. Waardt, W. Schairer, H. Rohde, G. Lehmann, E. S. R. Sikora, Y. R. Zhou, A. Lord, and D. Payne, "Field trial of 160Gbit/s OTDM add/drop node in a link of 275 km deployed fiber", in *Optical Fiber Communication Conference* (Optical Society of America, Los Angeles, 2004), pp. PDP1.

42 L. Krainer, R. Paschotta, G. J. Spühler, I. Klimov, C. Y. Teisset, K. J. Weingarten, and U. Keller, "Tunable picosecond pulse-generating laser with a repetition rate exceeding 10 GHz", *Electron. Lett.* **38** (2002) 225-227.

43 G. J. Spühler, L. Krainer, I. Kilburn, M. Dymott, K. J. Weingarten, R. Paschotta, and U. Keller, "Picosecond pulse-generating laser with repetition rate of 10 GHz and rms timing jitter <100 fs fully tunable over C-band", in *Optical Fiber Communication Conference and Exposition*, edited by J. E. Berthold and T. S. Afferton (Atlanta, U.S.A., 2003).

44 H. Tsuchida, "Pulse timing stabilization of a mode-locked Cr:LiSAF laser", *Opt. Lett.* **24** (1999) 1641-1643.

45 G. J. Spühler, M. Dymott, I. Klimov, G. Luntz, L. Baraldi, I. Kilburn, P. Crosby, S. Thomas, O. Zehnder, C. Y. Teisset, M. Brownell, K. J. Weingarten, R. Dangel, B. J. Offrein, G. L. Bona, O. Boccafusca, Y. Kaneko, L. Krainer, R. Paschotta, and U. Keller, "40 GHz pulse generating source with less than 350 fs timing jitter", *Electron. Lett.* **38** (2002) 1031-1033.

46 D. von der Linde, "Characterization of the noise in continuously operating mode-locked lasers", *Appl. Phys. B* **39** (1986) 201-217.

47 R. Paschotta, "Noise of mode-locked lasers. Part I: Numerical model", *Appl. Phys. B* **79** (2004) 153-162.

48 C. M. DePriest, T. Yilmaz, P. J. Delfyett, S. Etemad, A. Braun, and J. Abeles, "Ultralow noise and supermode suppression in an actively mode-locked external-cavity semiconductor diode ring laser", *Opt. Lett.* **27** (2002) 719-721.

49 C. Erny, G. J. Spühler, L. Krainer, R. Paschotta, K. J. Weingarten, and U. Keller, "Simple repetition rate tunable picosecond pulse-generating 10 GHz laser", *Electron. Lett.* **40** (2004) 877-878.

50 S. C. Zeller, L. Krainer, G. J. Spühler, K. J. Weingarten, R. Paschotta, and U. Keller, "Passively modelocked 40-GHz Er:Yb:glass laser", *Appl. Phys. B* **76** (2003) 1181-1182.

51 H. Takahashi, I. Nishi, and Y. Hibino, "10 GHz spacing optical frequency-division multiplexer based on arrayed waveguide grating", *Electron. Lett.* **28** (1992) 380-382.

52 H. Takara, E. Yamada, T. Ohara, K. Sato, K. Jinguji, Y. Inoue, T. Shibata, and T. Morioka, "106 x 10Gb/s, 25 GHz spaced, 640 km DWDM transmission employing a single super-continuum multi-carrier source", in *Conference on Lasers and Electro-Optics CLEO '01* (OSA, Baltimore, 2001), pp. Paper CPD11.

53 G. J. Spühler, P. S. Golding, L. Krainer, I. J. Kilburn, P. A. Crosby, M. Brownell, K. J. Weingarten, R. Paschotta, M. Haiml, R. Grange, and U. Keller, "Novel multi-wavelength source with 25-GHz channel spacing tunable over the C-band", *Electron. Lett.* **39** (2003) 778-80.

54 S. C. Zeller, L. Krainer, G. J. Spühler, R. Paschotta, M. Golling, D. Ebling, K. J. Weingarten, and U. Keller, "Passively mode-locked 50-GHz Er:Yb:glass laser", *Electron. Lett.* **40** (2004) 875-876.

International Journal of High Speed Electronics and Systems
Vol. 15, No. 3 (2005) 513–523

INTEGRATED WIDE-BAND CMOS DUOBINARY TRANSMITTER FOR OPTICAL COMMUNICATION SYSTEMS

RUI TAO

Institute of Electrical and Optical Communication Engineering, University of Stuttgart,
Pfaffenwaldring 47, D-70569 Stuttgart, Germany
Tel: +49 711 6857919, Fax: +49 711 6857900
taorui@int.uni-stuttgart.de

MANFRED BERROTH

Institute of Electrical and Optical Communication Engineering, University of Stuttgart,
Pfaffenwaldring 47, D-70569 Stuttgart, Germany
Tel: +49 711 6857919, Fax: +49 711 6857900
taorui@int.uni-stuttgart.de

Inter symbol interference (ISI) caused by dispersion in optical fibers is the major limiting factor on the achievable data rate or transmission distance in high-speed fiber-optic interconnects[1]. Compared with optical-domain and other electrical-domain dispersion compensation methods, duobinary coding technology is used due to its higher spectral efficiency. In this paper, a monolithically integrated duobinary optical transmitter has be designed and fabricated in standard 0.18 μm CMOS technology. This transmitter consists of a duobinary encoder and a VCSEL driver. The master-slave D flip-flop structure is used to generate the wideband one bit delay which is important for the realization of duobinary signal. An open drain structure is used in VCSEL driver design to provide the higher current output in low supply voltage condition. This duobinary transmitter can operate up to 5 Gb/s with 20 mA current output. The current consumption is only 50 mA under 1.8 V supply voltage.

Keywords: Dispersion; inter symbol interference; compensation; duobinary; encoding; transmitter; driver; VCSEL; eye-diagram.

1. Introduction

The recent popularity of the internet and multimedia communications has greatly increased the demands for large capacity optical communication systems. Due to the cost consideration, it is a challenge to effectively make use of the huge installed fiber in higher speed optical communication systems without degrading the repeater spacing and transport span between the nodes.

Inter symbol interference (ISI) is a fundamental problem in digital communications such as LANs or optical backplanes, where the multimode fibers are wildly used in such systems. In multimode optical fiber interconnects, for a given data rate, the ISI effectively sets the limit on the achievable link distance, due to the fact that it increases exponentially with the distance and thus, dominates the other penalties in the link power budget. Any further increase in the data rate would imply shorter distances.

The main source of ISI in a fiber-optic system is signal pulse broadening due to fiber dispersion. There are three types of dispersion in a fiber-optic system: modal dispersion, chromatic dispersion, and polarization mode dispersion[2]. In a multimode fiber, different mode groups have different velocities, which are called modal dispersion. Chromatic dispersion is due to the fact that different wavelengths of light have different velocities. The polarization-mode dispersion, which comes from different velocities of different polarizations, can be neglected in multimode fibers. The dispersion and the nonlinear effects can limit achievable data rate or transmission distance of optical pulses. Then, there is a trade off between the data bit rate and the transmission distance without the optical repeater.

In order to reduce the dispersion effect, different methods have been proposed and implemented in all three parts of a regenerator span (transmitter, fiber, and receiver), and in both optical and electrical domain[3-5]. The main criteria for a good dispersion reduction method are small power penalty (low loss), good integration with current networks, low cost, and adaptability.

Because of its spectral efficiency, the duobinary coding technology is an effective method in expanding the tolerable dispersion range. The duobinary optical transmission becomes a good candidate to solve the problem. The higher bit rate data can be transferred in the installed fiber because of the compact spectrum of duobinary signal. Therefore, there is a strong demand for an integrated duobinary transmitter and receiver in high-speed fiber-optic communications, especially in cost-sensitive short-haul systems. Several Gbit/s hybrid integrated duobinary optical transmission systems have been demonstrated[6-7].

A duobinary optical transmitter, which consists of a duobinary encoder and a laser diode driver, is a critical part of practical optical duobinary transmission systems. In this paper, a 5 Gb/s fully differential duobinary CMOS transmitter is designed, fabricated in advanced 0.18 μm CMOS technology.

This paper is organized as follows. In Section 2, the duobinary transmission theory is introduced and compared with the general binary transmission technique. In Section 3, the design of the duobinary transmitter is described. In Section 4, the IC fabrication aspects are discussed. The measurement results are presented in Section 5. Finally, in Section 6, conclusions are given.

2. Duobinary Theory

In the optical system, an effective channel (fiber) bandwidth limitation resulting from fiber dispersion limits the achievable maximum transmission distance or highest data rate. It shows theoretically that with a zero chirp transmitter, the system penalty duo to the fiber dispersion is set by the Fourier bandwidth of the data. A standard dispersion limited bit rate distance product is give by:

$$B \cdot L \leq \frac{1}{4D\sigma_\lambda}$$

B is the data bit rate, L is propagation distance, D is the fiber dispersion coefficient, and σ_λ is the Fourier bandwidth of the signal. It can be seen that the bit rate distance product is improved if the width of σ_λ can be reduced. Duobinary coding technology is used duo to its higher spectrum efficiency.

Duobinary transmission technique is a special case of the multilevel transmission technique, which was presented by Dr. Adam Lender in 1960s[8]. It is well known that the duobinary technique is used to double the transmission capacity (speed) at a given channel bandwidth.

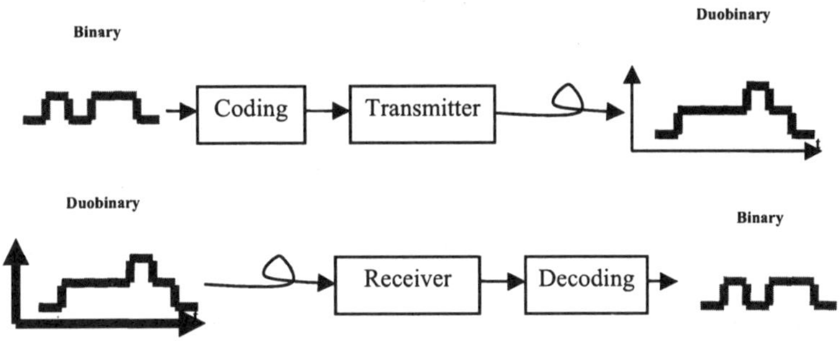

Fig. 1. System block of duobinary transmission system

The system block of the duobinary transmission system is shown in Fig. 1. The original electrical binary signal is encoded and converted to the duobinary optical signal. The duobinary optical signal is transferred in the optical fiber. At the receiver side, after the optoelectronic conversion, the duobinary signal is decoded to get the original binary signal.

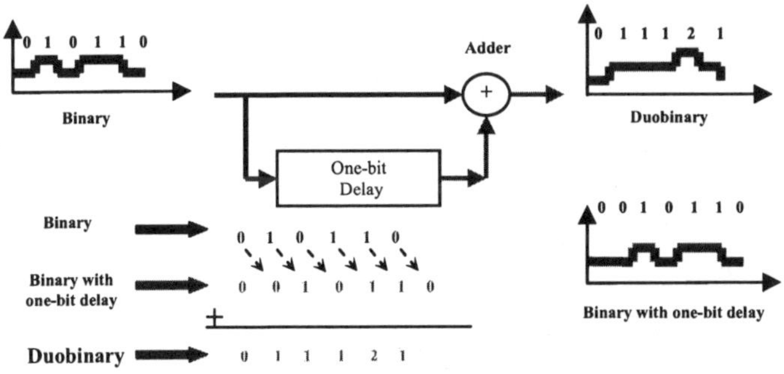

Fig. 2. The concept of idea duobinary encoding.

As the critical part of duobinary transmitter, an idea duobinary encoder is show in Fig. 2. A standard electrical binary NRZ data signal is delay 180° in phase (one bit delay). The sum is obtained by an ordinary addition of the signals from two paths, one a direct connection and the other through a one bit delay cell. At the output of adder, the three levels duobinary signal is obtained.

The amplitude of overall transfer function of this idea duobinary encoder is derived as:

$$|H(f)| = |\cos \pi f T| \tag{1}$$

The power spectrum density functions of original binary and duobinary signal derived from this binary signal are shown in Eq. (2) and (3):

$$S_b(f) = \frac{T}{4}\left(\frac{\sin \pi f T}{\pi f T}\right)^2 \qquad \text{Binary} \tag{2}$$

$$S_d(f) = S_b(f) \cdot |H(f)|^2 = T\left(\frac{\sin 2\pi f T}{2\pi f T}\right)^2 \qquad \text{Duobinary} \tag{3}$$

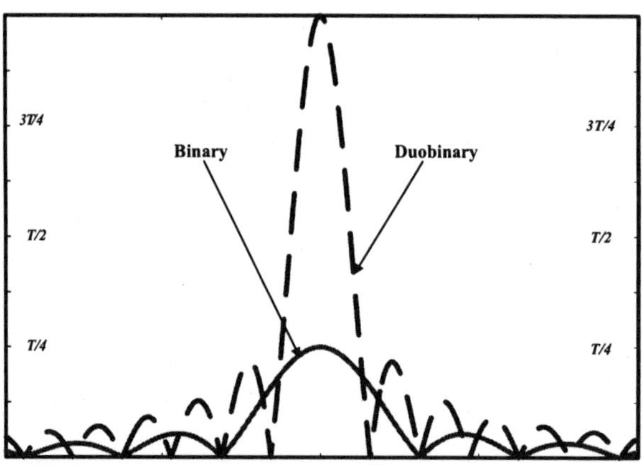

Fig. 3. The power spectrum density function of binary and duobinary signal (Dash line is duobinary signal, solid line is binary signal)

Fig. 3 shows the power spectrum density function curves of two equations. Compared to the normal binary NRZ signal, the derived duobinary signal occupies only half the signal bandwidth.

The duobinary technique effectively reduces the width of σ_λ by a factor of 2. As a result, the maximum transmission distance or highest data bit rate is doubled at the expense of a small amount of signal processing after detection to restore the original binary signal. A further advantage is that the additional circuitry required to achieve duobinary operation is simple, but the technique requires many of the electrical and optoelectronic components to have only half of the electrical bandwidth compared with standard binary signal. Reductions of cost and complexity are thus provided.

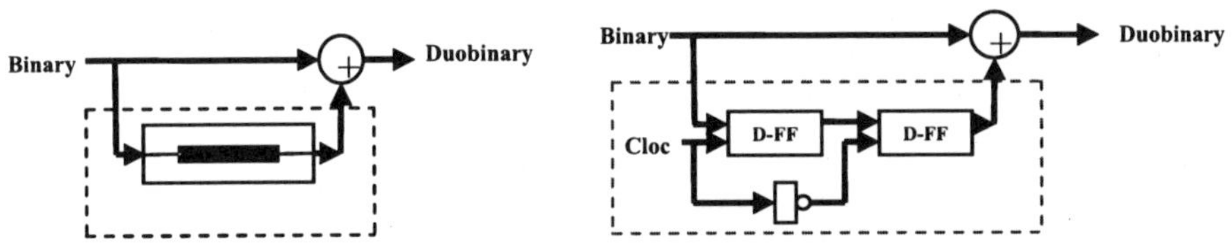

(a) Microstripe structure (b) D-flipflop structure

Fig. 4. Two kinds of one bit delay cell

One bit delay cell is the indispensable part of duobinary encoder. There are two methods which can be used to generated one bit delay: microstripe structure and D-flipflop structure which are shown in Fig.4 (a) and (b).

In microstripe structure, the phase delay is depended on the operating frequency and the physical parameter of microstripe line, so the disadvantages of this one bit delay cell are narrow operating bandwidth and large chip area. It can only work at certain frequency to generate one bit delay.

Compare to microstripe structure, a master-slave D-type flip-flop (MS-DFF) is used to achieve a one bit delay time, which allows both delay time adjustment by shifting the clock time for the MS-DFF and bit rate flexible operation. So the one bit delay of D flip-flop structure is controlled by the clock signals, it can work in larger bandwidth.

3. System and Circuits

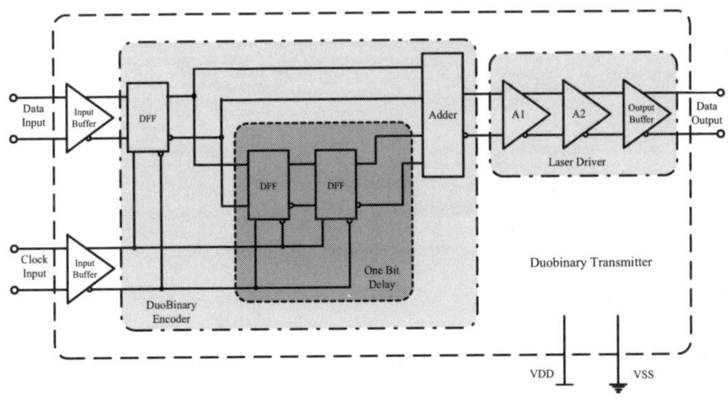

Fig. 5. The system block of duobinary transmitter

Fig. 5 is the system block diagram of the whole duobinary transmitter, it consists of a data input buffer, a clock input buffer, a duobinary encoder, and a VCSEL driver. Every part will be described in detail.

3.1. *Input buffer*

The input buffer is used to provide the 50 Ω input impedance matching, adjust the DC level of input signal, and amplify the data and clock signals to reshape the waveforms.

3.2. *Encoder*

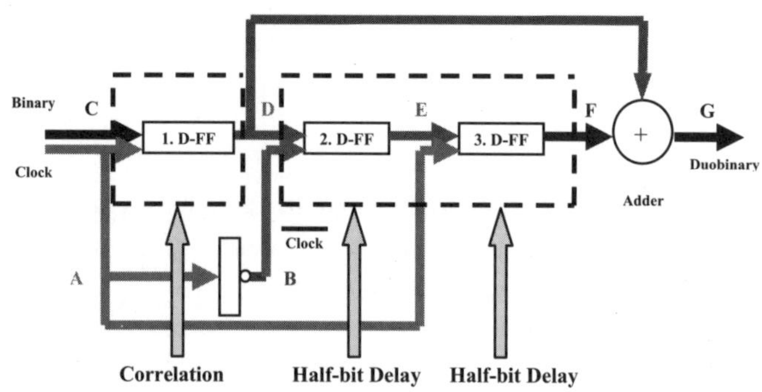

Fig. 6. Duobinary Encoder

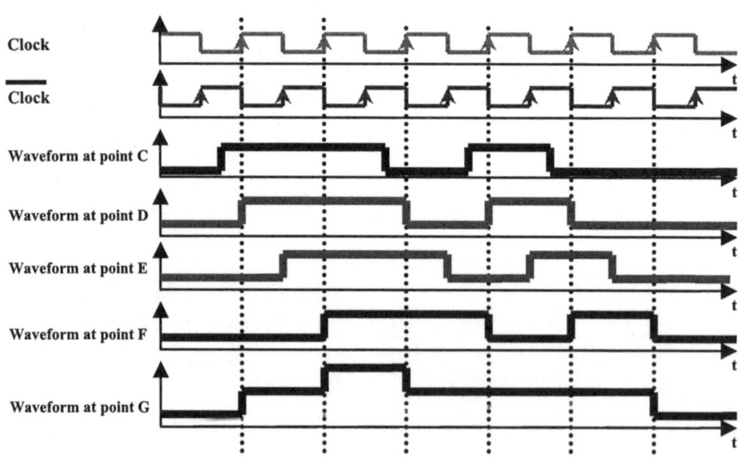

Fig. 7. Waveform output at every nodes of duobinary encoder

The Fig. 6 is the system block of the duobinary encoder. The input standard binary signal is encoded by this encoder and converted to a three level duobinary encoded signal. Fig.7 is the waveform output at the different point of this encoder. It is used to explain how this duobinary encoder works. This encoder consists of a D-FF for retiming, a wide bandwidth MS D flip-flop structure one-bit delay cell and an adder to generate the duobinary signal.

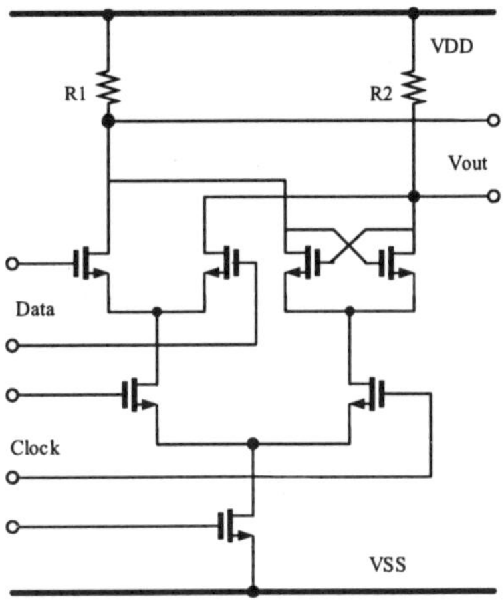

(a) Schematic of D-latch

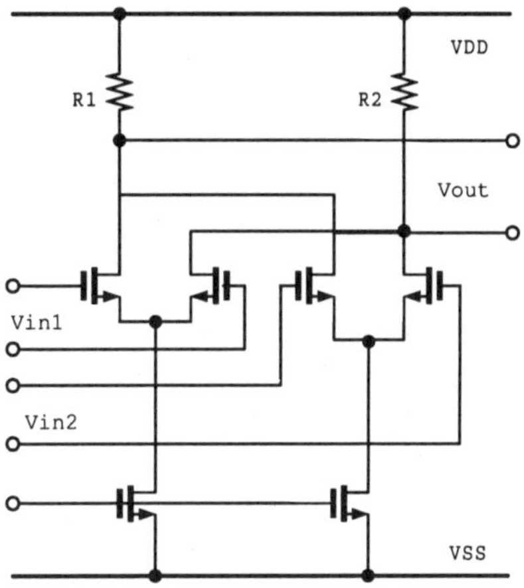

(b) Schematic of adder

Fig. 8. Schematic of D-latch and adder

The input binary data signal C is retimed and reshaped after the first D flip-flop. After this D-FF, the relationship between the data and clock signal is set up. Then, the retimed signal D is delayed 180° degree in phase by a master-slave DFF. The retimed signal D and delayed signal F is added by an adder to get a three level duobinary encoded signal G.

The basic cell for the D-Latch is show in the Fig.8 (a). The gate width of the read and latch pair, and the gate width ratio between the read and latch pair are optimized to get the high operating speed. The D flip-flop is realized by cascading these two D latches.

The schematic of adder is shown in Fig.8 (b). It is use to add the original signal and the signal with one bit delay to get the duobinary signal. Also, this adder can be treated as a low pass filter. This bandwidth limiting of this LPF suppresses the side lobes of the signal spectra and expands the dispersion tolerance.

3.3. *Driver design*

The three level duobinary encoded signal is needed to be converted to an optical duobinary signal. A laser/modulator driver must be used to provide the high voltage or current output to driver the external laser diode or MZ modulator. In our case, the whole transmitter is monolithically integrated using 0.18 μm CMOS technology, the supply voltage is only 1.8 V. It is difficult to realize a MZ modulator driver whose output should be about 4-5 V. Also due to the cost consideration, the VCSEL diodes are widely used now. Numerous groups have demonstrated large signal modulation of VCSEL up to 12.5 Gb/s[9-10].

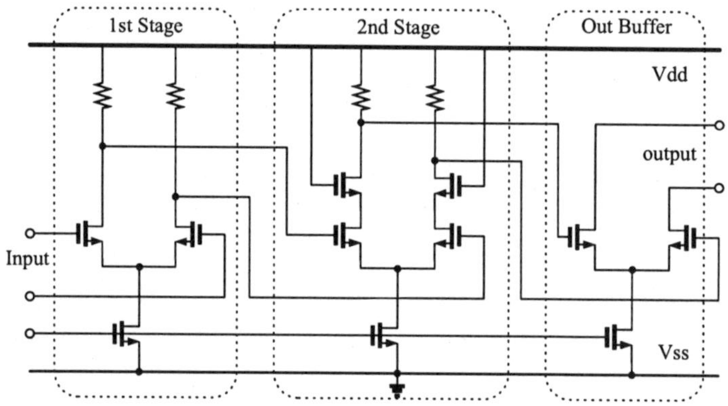

Fig. 9. The block of VCSEL diode driver

So in this transmitter, a VCSEL diode driver with three level current outputs is designed. The circuit topology is showed in Fig. 9. The first common source differential amplifier is used to provide the high gain. For the second stage, a common source common gate differential amplifier is used. This cascade amplifier offers high voltage gain while minimizing the extra Miller capacitance caused by the amplifier gain and FET gain-drain capacitance in order to broaden the bandwidth. The low power supply voltage is not sufficient for the laser driver stage, so the open drain structure is used in the output buffer to provide the high current to the VCSEL diode. Another reason is that the problem of undershoot in the laser driving signal caused by the impedance mismatch between the driver IC and the VCSEL laser is solved by using the open drain structure. The dimensions of transistors in the output buffer stage are determined by the maximum modulation current of the laser diode[11].

4. IC Fabrication

The IC was fabricated using 0.18 μm gate length, single poly, six-metal, triple well standard CMOS technology by means of the CMP program, France. The transit frequency f_T of this very deep sub-micron CMOS technology is about 40 GHz. The microphotograph of duobinary transmitter IC is shown in Fig. 10. The chip area is about 0.55x0.6 mm2.

Several criteria for this high frequency laser driver layout are considered to minimize the possible parasitic elements and interference. The differential input and output signals with a GSSG configuration are on opposite sides of the chip to allow straight-through connection between the stages. Layout was carried out preserving the symmetry of the differential amplifier to reduce offset from layout and processing. Bias, power line and power pads of the small signal and the large signal paths are separated to prevent interference. Input and output are arranged as far as possible to get the signal purity. Due to the large current output, double guard rings are used to minimize the possibility of latch-up.

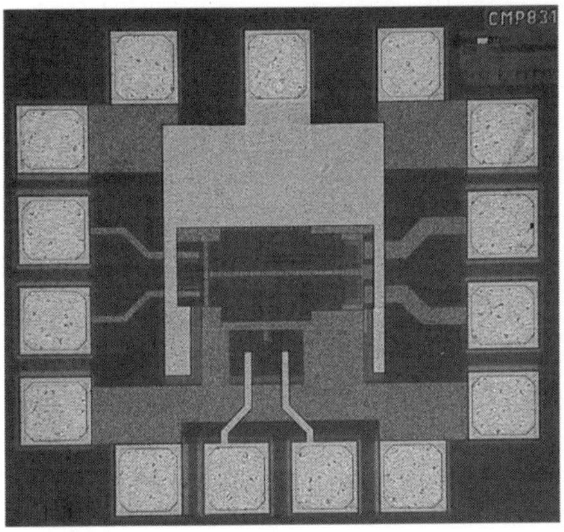

Fig. 10. Micrograph of CMOS duobinary transmitter

5. Tested Results

5.1. *On wafer measurement*

This duobinary transmitter was tested on wafer. The experimental setup is shown in Fig. 11. The data input is the PRBS signal from the pseudorandom pattern generator (PPG). Due to the open drain structure of laser diode drive, two bias Tees are use to give the DC bias to the output buffer. The input impedance of the VCSEL is about 50 Ω, which is same as the input impedance of high speed oscilloscope.

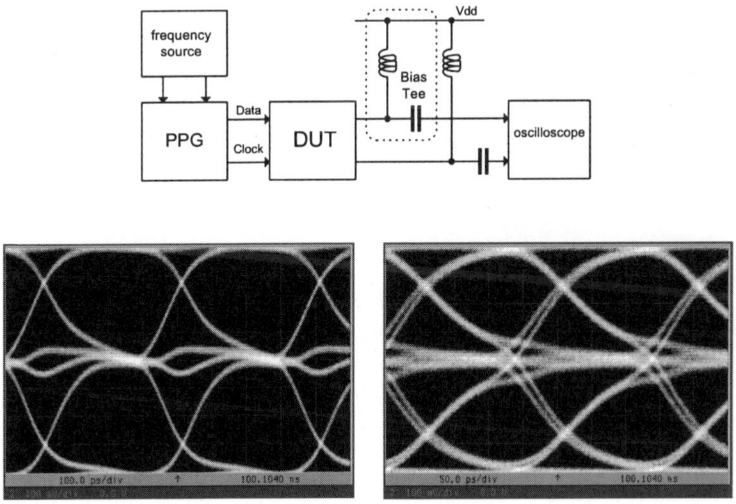

Fig. 11. Measurement setup and the duobinary output eye diagram with 2.5 Gb/s (left) and 5 Gb/s (right) binary PRBS input

The output eye diagram with 2.5 Gb/s signal input and 5 Gb/s signal input were shown in Fig. 11. Clear eye openings with short rise and fall time were obtained. The amplitude of output voltage signal is 1 Vp-p at 50 Ω load, corresponding to the output current 20 mA which is high enough to drive the VCSEL diode.

The phase of clock signal fed into the chip can be adjusted in order to ensure accurate one-bit delay circuit operation. The power consumption is 50 mA at 1.8 V supply voltage.

5.2. *System measurement*

After the on-wafer measurement, this duobinary transmitter was testes in the real optical transmission system. The experimental setup is shown in Fig. 12.

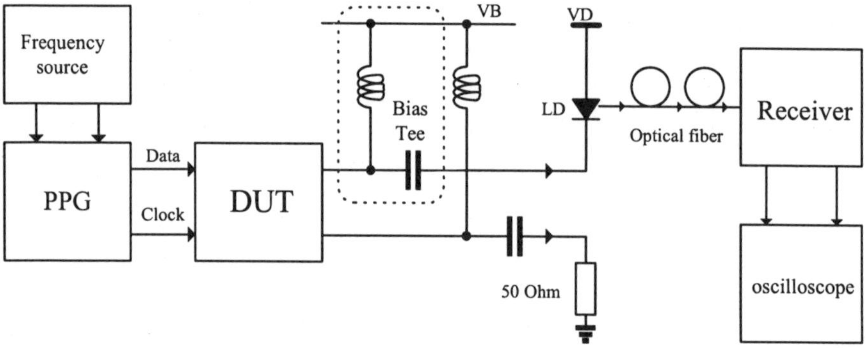

Fig. 12. System Measurement setup

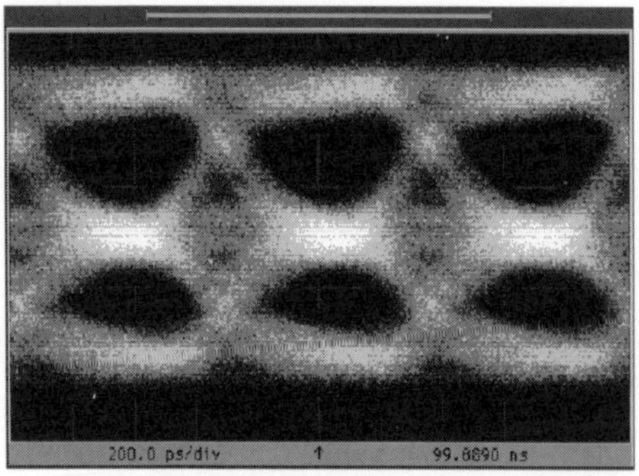

Fig. 13. 1.5 Gb/s output eye diagram of system measurement

One output of the duobinary transmitter is connected to a laser diode, another output is terminated by a 50 Ω load. The generated optical signal is coupled and transmitted through a long fiber. A high speed optical receiver is used converts the optical signal into voltage one which will be measured using high speed oscilloscope. The detected output eye diagram is shown in Fig. 13. Due to the bandwidth limitation of laser driver and optical receiver, the clean output eye diagram was get at 1.5 Gb/s.

6. Conclusion

A duobinary transmitter IC has been designed and realized using 0.18 μm standard CMOS technology. Error free operation of this circuit was confirmed with 5 Gb/s 2^{31}-1 PRBS signal. The power consumption is only 90 mW under 1.8 V supply voltage. With this duobinary transmitter, the system transmission capacity can be doubled even using the installed the optical fiber systems.

Reference

1. Hui Wu, Jose A. Tierno, et al, "Integrated Transversal Equalizers in High-Speed Fiber-Optic Systems", IEEE JOURNAL OF SOLID-STATE CIRCUITS, VOL. 38, NO. 12, DECEMBER 2003, pp. 2131-2137.
2. G. Agrawal, Fiber-Optic Communication Systems, 3rd ed. New York: Wiley, 2002.
3. J. Winters and R. Gitlin, "Electrical signal processing techniques in long-haul fiber optics systems," IEEE Trans. Commun., vol. 38, pp. 1439–53, Sept. 1990.
4. J.Winters, R. Giltin, and S. Kasturia, "Reducing the effects of transmission impairments in digital fiber optic systems," IEEE Commun. Mag., vol. 31, pp. 68–76, June 1993.
5. B. Jopson and A. Gnauck, "Dispersion compensation for optical fiber systems," IEEE Commun. Mag., vol. 33, pp. 96–102, June 1995.
6. X.Gu, and L.C.Blank, "10 Gbit/s unrepeated three-level optical transmission over 100 km of standard fiber" Electronics Letters, 9th December 1993, Vol.29, No.25, pp2209-2211.
7. A. Matsuura, K. Yonenaqge, et al, "High-speed transmission system based on optical odified duobinary signals" Electronics Letters, 29th April 1999, Vol.35, No.9, pp736-737.
8. A.Lender, "Correlative digital communication techniques" IEEE Trans Commun. Technol., Vol. Com-12, 1964, pp128-135.
9. M. Kircherer, R.Jager, R.King, et al, "Single- and Multi-Mode VCSELs for 12.5 Gb/s Data Links", In Lasers and Electro-Optics Europe, 2000 conference Digest paper CTuG2.
10. M.Webster, R.V.Penty, et al, "Performance of 10 Gb/s VCSEL operating at 85C", In conference of Lasers and Electro-Optics (CLEO) 2000, pp.201-202.
11. Rui Tao, Manfred Berroth, Zhigong Wang, "Low Power 10 Gb/s VCSEL Driver for Optical Interconnect", Electronics Letters, vol. 39, No. 24, November 27th, 2003, pp. 1743-1744.

International Journal of High Speed Electronics and Systems
Vol. 15, No. 3 (2005) 525–548

World Scientific
www.worldscientific.com

A 10 GB/S EQUALIZER WITH INTEGRATED CLOCK AND DATA RECOVERY FOR OPTICAL COMMUNICATION SYSTEMS

D. S. MCPHERSON, H. TRAN, AND P. POPESCU

Quake Technologies Inc., 80 Hines Road,
Ottawa, Ontario, K2K 2T8, Canada
douglasm@mail.quaketech.com

A 10 Gb/s analog continuous-time equalizer with integrated clock and data recovery circuit is presented. It is designed to recover signals degraded by chromatic and polarization mode dispersion. The key components in the design are a feedforward equalizer and a decision feedback equalizer, the parameters of which are electronically adjustable. Both circuit blocks are fully described and characterized with emphasis on minimizing self-induced distortion and maximizing high-speed performance. In addition to the equalizer and the clock and data recovery, the circuit also includes an integrated automatic gain control. The circuit is implemented in a commercial 0.18 μm SiGe BiCMOS technology and consumes 900 mW. The capacity of the equalizer to mitigate signal impairments is demonstrated using three electrically generated channels.

Keywords: equalizers; optical fiber dispersion; optical communications.

1. Introduction

With the growth of high-bandwidth applications at the premises and metro level, 10 Gb/s fiber optic links provide an attractive solution for expanding capacity in metropolitan area networks (MANs). This presents a significant challenge because 10 Gb/s signals are much more susceptible to the impairments introduced by the installed fiber, namely chromatic and polarization mode dispersion. Although passive optical solutions exist for 10 Gb/s long-haul networks, issues of cost and suitability prohibit their use in the aforementioned shorter reach applications[1]. For this reason, electronic techniques, in particular adaptive equalizers, have attracted a considerable amount of interest lately[2-5]. The principal advantage of adaptive equalizers, besides low cost, is their capacity to mitigate fiber impairments without foreknowledge of the channel. Moreover, unlike passive optical techniques, they can track changes in the channel, such as those introduced by temperature, mechanical stress and aging.

From an implementation perspective, equalizers can be broadly divided into three categories by the way that they operate and the demands that they place on the available semiconductor technology. Of these, a fully digital implementation employing maximum likelihood sequence estimation (MLSE) offers the highest theoretical level of performance[6,7]. However, such an approach is costly, not only in terms of power consumption, but in chip area and complexity as well. At 10 Gb/s data rates, the necessary electronics, both for analog to digital conversion (ADC) and for subsequent processing, are state-of-the-art[8] and some form of interleaving is often unavoidable[9]. An alternative is to use a discrete time approach where the input signal is sampled at greater than 10 GSamples/s using a track and hold amplifier[10,11]. Subsequent processing of the signal is then carried out using less demanding analog techniques[12]. Although 10 Gb/s track and holds and ADCs exist, their power and size requirements are such that they are of limited commercial appeal at this time.

The implementation presented in this paper uses a continuous time approach wherein the signal is processed through the equalizer circuitry without any sampling. The design consists of an automatic gain control (AGC) followed by a feedforward equalizer (FFE) and a decision feedback equalizer (DFE). All of these blocks use analog circuitry, such as four-quadrant multipliers and delay lines to mitigate the fiber-induced impairments. To provide an additional performance enhancement, the equalizer is integrated with a clock and data recovery (CDR). The main advantage of the analog continuous time approach is that it is less demanding on the technology, relatively small, and consumes less power than the discrete time or completely digital implementations. To realize these advantages at 10 Gb/s, innovative high-speed circuits and/or implementations of existing circuit topologies are required. These innovations form the basis of this paper and are discussed in detail. Some of them are also covered in several pending patents[13-15].

The present equalizer is designed to mitigate dispersion impairments arising in extended reach single-mode (80-100 km) fiber links at bit-rates in the range of 10.7 to 11.1 Gb/s[16]. The parameters of the equalizer are adjusted electronically using an external microcontroller connected to the IC via a 2-wire industry standard interface.

2. Circuit design

A block diagram of the equalizer with clock and data recovery is shown in Fig. 1. The device supports an input signal in the range of 10 to 600 mV_{pp}/side, which is adjusted to a fixed level by an automatic gain control, processed through a feed-forward equalizer and applied to a decision circuit (DEC). A decision feedback equalizer is also included. Combined, these elements can address both pre- and post-cursor intersymbol interference (ISI). After the DEC, the signal is presented to the CDR as well as a test port monitor (EQU OUT).

In many cases, equalizers produce an output eye that exhibits significant jitter and closure, as well as horizontal and vertical asymmetry. If an external re-timing device is used to recover the data, these characteristics are further compounded by bandwidth limitations and distortion associated with I/O circuitry, PCB interconnect, and integrated circuit (IC) package parasitics. By integrating the CDR directly with the equalizer, these latter effects can be averted, and consequently, the receiver performance improved. This architecture, of which the present design is the first reported example for 10 Gb/s fiber optic links, also has the advantage that it relaxes the requirements for the subsequent

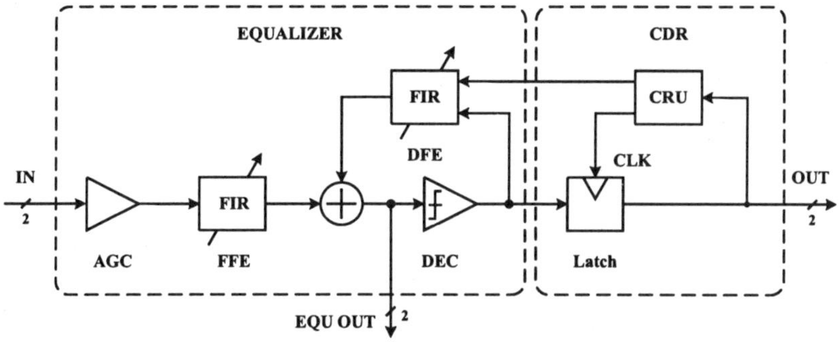

Fig. 1. Block diagram of the equalizer with integrated CDR.

demultiplexer, both in terms of its front-end design and the locking capability of its CDR. Moreover, it represents the first step towards integrating the equalizer with a serializer/deserializer (SERDES).

2.1. *Automatic gain control*

Laser output power, detector sensitivity, span length, connector losses, and optical amplification are just a few of the many factors that lead to significant variability in the transmission characteristics of fiber optic links. In order to ensure that such signals are within the range of levels that can be accommodated by the equalizer, an AGC is necessary. A block diagram of the circuit implemented in the present IC is shown in Fig. 2. It consists of three stages and has a dynamic range of 35 dB. The two variable gain stages, which are based on a wide-band analog multiplier topology[17-19], are controlled automatically by a digital on-chip adaptation algorithm. Given the potential for high gain values when both G1 and G2 are near their maximum, the AGC also incorporates an autozero DC cancellation circuit to remove any offsets that may arise due to device mismatch.

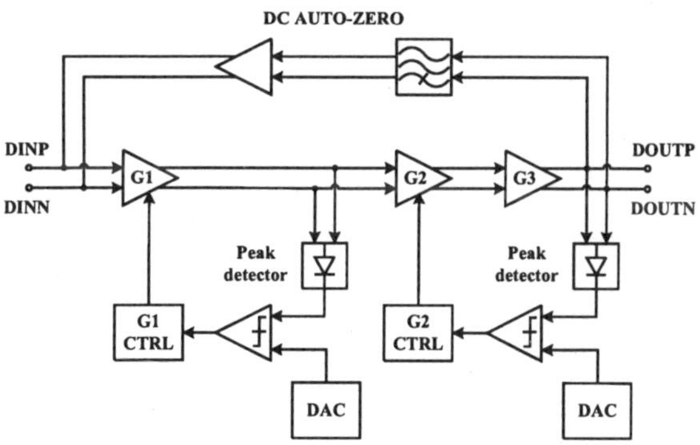

Fig. 2. Automatic gain control circuit with autozero DC feedback cancellation circuit.

2.2. *Feedforward equalizer*

The function of a feedforward equalizer is to assume the inverse characteristic of the dispersive channel. In so doing, it restores the higher-frequency signal components and substantially reduces the effects of ISI. The present design uses a fractionally-spaced and fully differential transversal filter to process the received signal and an external microcontroller to adjust its tap weight coefficients. These coefficients are stored in registers, which in turn control the settings on a set of seven 7-bit digital to analog converters (DACs).

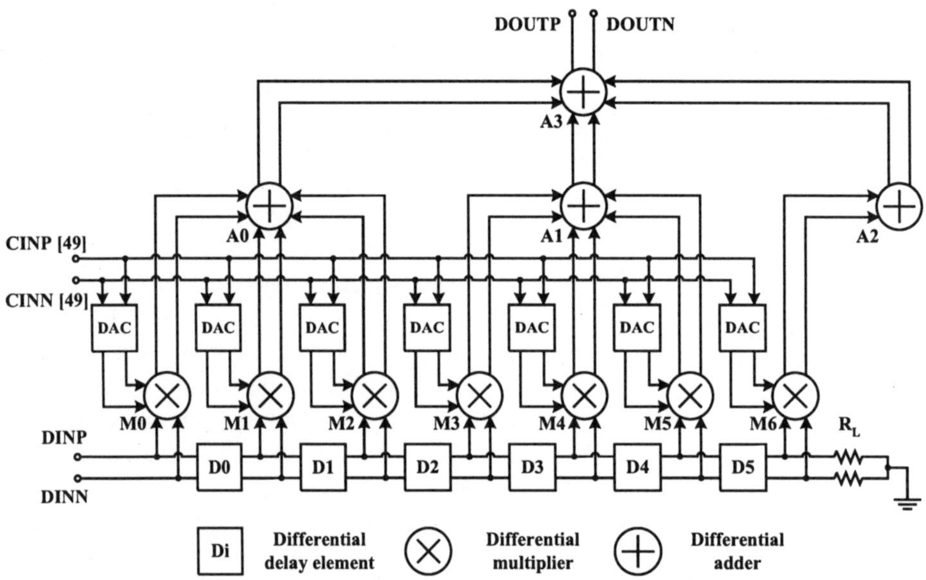

Fig. 3. Block diagram of the feedforward equalizer.

A block diagram of the transversal filter is shown in Fig. 3 along with the DACs. It is based on a hierarchical multiplier-adder implementation rather than a distributed amplifier configuration[20-21]. Although the latter topology has the advantage that it offers a potential boost in bandwidth by the absorption of the parasitic input/output capacitances into the artificial transmission lines, it has some constraints. Firstly, in order to obtain the bandwidth enhancement, the inductive component of the delay elements must combine with the capacitances to form the desired artificial transmission line characteristic impedance. This stipulation imposes an upper bound on the selection of the delay length per tap. Secondly, having a low output impedance requires that the taps swing significant amounts of current in order to generate an appreciable signal amplitude. By comparison, the hierarchical multiplier-adder implementation has no constraints on the tap spacing at the input, nor any constraints on the output impedance.

2.2.1. *Delay elements*

One of the principal impediments to the realization of a purely analog feed-forward equalizer is the provision of sufficiently wide-bandwidth delay structures. For circuits operated at data rates of 10 Gb/s or more and having in excess of 2-3 taps, the insertion loss, reduction in bandwidth and group delay variation from the first to the last tap can significantly impair the performance of the equalizer. To minimize this impairment, the delay structures in the present design are implemented as loosely coupled differential microstrip transmission lines. Compared to transistor or filter based implementations, this approach provides high bandwidth, zero power dissipation, lower group delay variation and less susceptibility to variations in component tolerances and temperature.

Although the advantages of using transmission lines as high-speed analog delay elements are self-evident, their actual implementation poses a significant challenge. This is particularly true for FFE circuits where the combined delay of all of the transmission line elements must exceed several multiples of the bit period. Unfortunately, delays of this order require transmission lines that are of significant physical length, a fact that can be a major detriment if the circuit is to be implemented using IC technology. In order to conserve chip real-estate, the conductors must be meandered and at the same time be quite narrow, with unfavorably low conductor thickness to width ratios (e.g. 1:2). At exceptionally long lengths, such transmission line geometries exhibit a range of characteristics that are not readily captured by commercially available computer aided design (CAD) models. For this reason, a 44 mm-long test structure of the proposed delay design was fabricated and characterized using a vector network analyzer (VNA).

To appreciate the impairments that transmission lines of this length and cross-sectional geometry exhibit, the characteristic resistance, inductance, conductance, and capacitance (*RLGC*) were extracted from the measured *S*-parameters. In so doing, it is also possible to scale the transmission line to other lengths while still preserving all of the impairments. Since the coupling between adjacent conductors is less than -25 dB, the differential pairs are treated as isolated single conductors. The first step in the extraction is to convert the 2-port *S*-parameters to *ABCD*-parameters. Having done so, the characteristic impedance Z_C and propagation constant γ can be readily obtained from

$$\gamma = \cosh^{-1}\left(A_{11}\right)/l \tag{1}$$

$$Z_C = \frac{A_{12}}{\sinh\left(\gamma l\right)} \tag{2}$$

where l is the physical length of the transmission line. Once obtained, these values can be used to determine the *RLGC* parameters of the transmission lines by means of the telegraphist's equations

$$Z_C = \sqrt{\frac{R + j\omega L}{G + j\omega C}} \tag{3}$$

$$\gamma = \sqrt{\left(R + j\omega L\right)\left(G + j\omega C\right)}\ . \tag{4}$$

The individual parameters are then obtained by combining the equations as follows

$$R = \mathrm{Re}(\gamma Z_C) \tag{5}$$

$$L = \mathrm{Im}(\gamma Z_C)/\omega \tag{6}$$

$$G = \mathrm{Re}(\gamma / Z_C) \tag{7}$$

$$C = \mathrm{Im}(\gamma / Z_C)/\omega \ . \tag{8}$$

Following this established extraction methodology[22], the *RLGC* parameters for the 44 mm-long test structure are readily determined and are plotted in Fig. 4. With respect to Fig. 4(a), the extracted resistance is 1.5 Ω/mm at DC. Over a length of 44 mm, this resistance results in a low-frequency input return loss of 8 dB and an insertion loss of 4.4 dB. Contrary to expectations, the skin-effect losses only increase as $\sqrt{f}$ up to 3 GHz, beyond which the relationship assumes a more linear behavior. As shown in Fig. 4(b), the variation in inductance with frequency is much more dramatic with a very sharp rise at low frequencies. Although it is not possible to give a definitive explanation for this

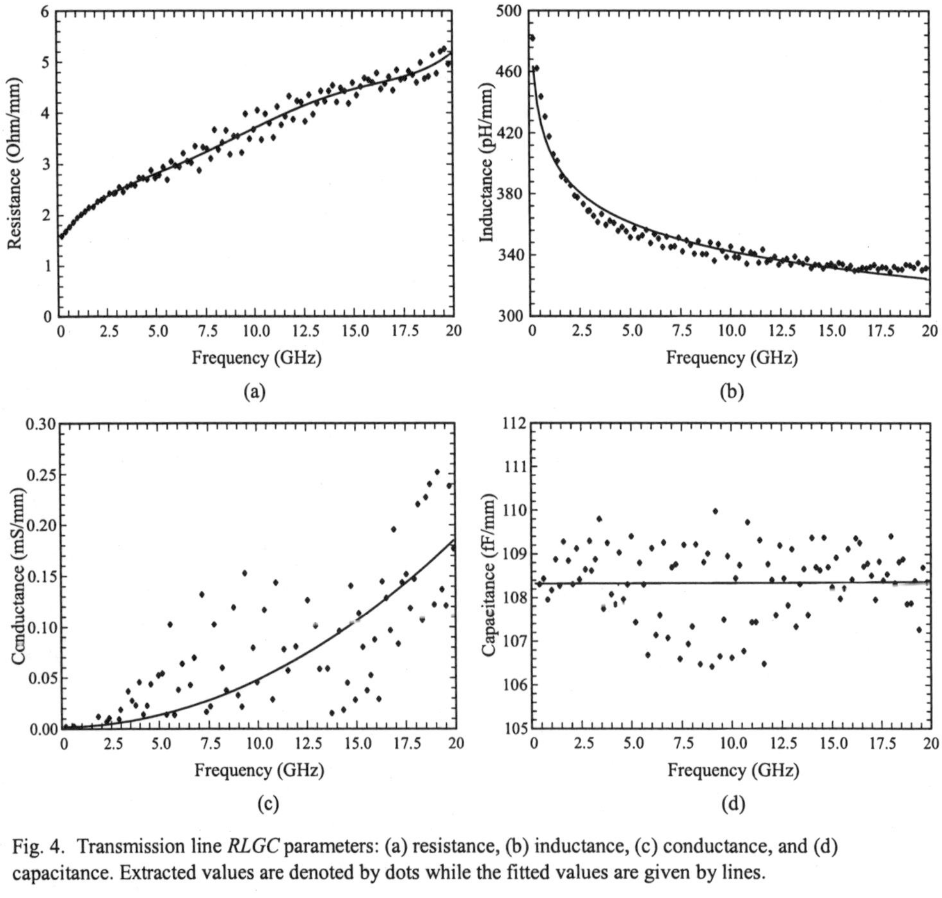

Fig. 4. Transmission line *RLGC* parameters: (a) resistance, (b) inductance, (c) conductance, and (d) capacitance. Extracted values are denoted by dots while the fitted values are given by lines.

behavior, the fact that the conductor is less than one skin-depth thick below 1 GHz necessarily implies that the current density will vary more rapidly at lower frequencies. Indeed, most conventional assumptions regarding conductor losses are premised on having a conductor strip thickness of at least three to four skin depths[23]. The conductance and capacitance curves, Figs. 4(c) and 4(d) show that the dielectric losses are slight in comparison to the conductor losses and vary significantly less with frequency.

In order to capture all of these observed impairments in the design of the FFE, smooth curves were fitted to the extracted *RLGC* parameters and are shown as lines in Fig. 4. The resulting curves can then be scaled to any desired length, converted back to *ABCD*-parameters by

$$[\mathbf{A}]=\begin{bmatrix} \cosh(\gamma l) & Z_C \sinh(\gamma l) \\ (1/Z_C)\sinh(\gamma l) & \cosh(\gamma l) \end{bmatrix} \tag{9}$$

and then converted to *S*-parameters. For more information on these parameter conversions, the reader is referred to R. E. Collin[24].

Fig. 5 shows the measured versus fitted *S*-parameters and group delay for the 44 mm long test structure. The results show very good agreement for both the input return loss and insertion loss. This includes the sharp rise in the input return loss at low frequency, which is the result of the reactive terms in equation (3) becoming vanishingly small while the series resistance term dominates. The group delay shown in Fig. 5(c) indicates that the sharp rise at low frequency is captured, although the agreement is not exact. This is because the group delay is very dependent on the exact shape of the inductance versus frequency characteristic.

Observing these curves, it is readily apparent that the delay elements will introduce a significant amount of frequency dependent loss and delay variation. Viewed as an eye diagram, the 10 Gb/s signal will be almost completely closed and will contain upwards or 40 ps of jitter based on the insertion loss and group delay results alone. The high input return loss at low frequency will add to the distortion in the eye resulting in a equalizer that adds more signal impairment than it actually removes.

The solution to these issues is based on the observations made above, particularly regarding the *R* and *L* variation. In order to blunt their effects at low frequency, it is proposed to increase the value of the conductance term *G* in equations (3) and (4). If chosen correctly, this can counterbalance the varying *R* and *L* terms at low frequency while having a minimal effect at higher frequencies. In order to increase the conductance term, shunt resistors are connected between the differential transmission line pairs at periodic intervals. A useful guide for selecting the value of the shunt resistors is to make the single-ended input return loss Z_{IN} equal to 50 Ω at low frequency. It can be shown that a cascade of *N* resistor *pi* networks has the following input impedance Z_{IN}, given a final load resistance of R_L

$$Z_{IN}=\frac{A_{11}^N R_L + A_{12}^N}{A_{21}^N R_L + A_{22}^N} \tag{10}$$

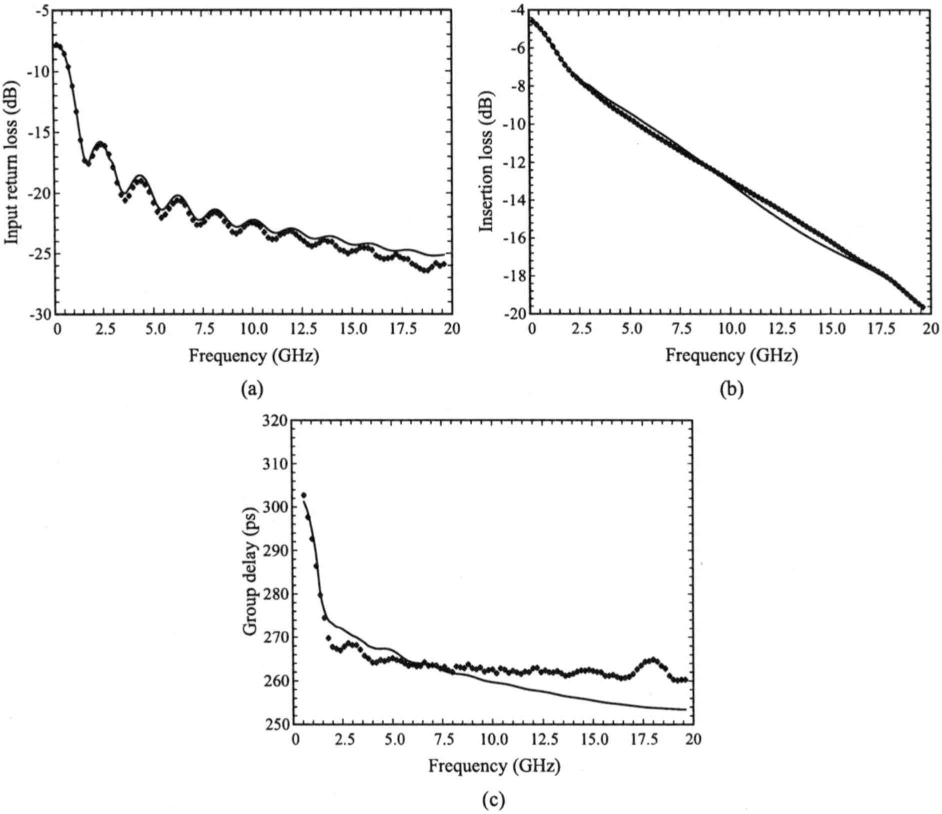

Fig. 5. Comparison of measured to fitted *S*-parameters: (a) input return loss, (b) insertion loss, and (c) group delay. Measurements are denoted by dots while the fitted values are given by lines.

where

$$[\mathbf{A}] = \begin{bmatrix} R_P(R_S + R_P) & R_S \\ (R_S + 2R_P)/R_P^2 & R_P(R_S + R_P) \end{bmatrix} \tag{11}$$

and R_P and R_S are the inserted shunt resistance and the transmission line DC series resistance, respectively. To account for the fact that the shunt resistance is duplicated at the input and output of each *pi* segment, the implemented value is actually half the size of R_P. In terms of placement, it would be ideal if the shunt resistance could be distributed over the entire length of the transmission line. However, it turns out to be almost as effective, and far more realistic, to insert two shunt resistors per FFE tap, one at the input to each tap and one in between taps. This is markedly superior to one resistor per tap and almost as effective as three or four resistors per tap.

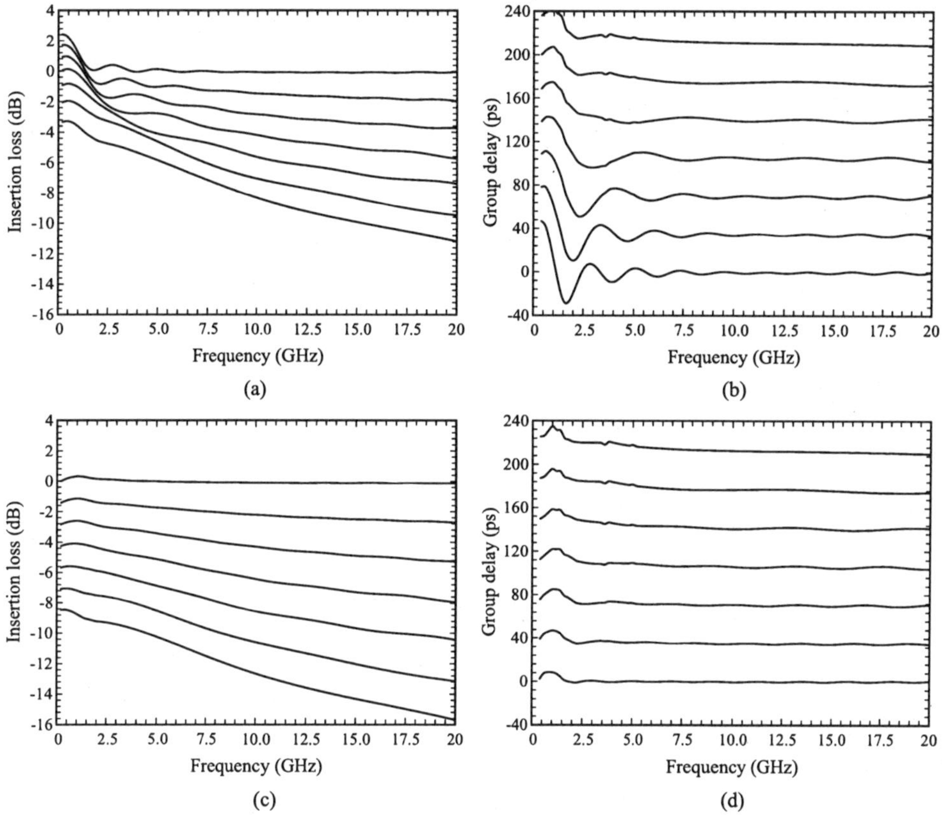

Fig. 6. Small-signal response at the equalizer's tap inputs: (a) uncompensated insertion loss, (b) uncompensated group delay variation, (c) compensated insertion loss, and (d) compensated group delay.

Fig. 6 illustrates the efficacy of the compensation scheme in reducing distortion introduced by the delay elements. In Fig. 6(a) and (b), the performance is particularly poor at the input tap. While this may not be readily apparent, given that the signal has not been processed along the transmission line, the input impedance seen looking towards the load is actually much worse at the first tap than it is at tap inputs closer to it. The compensated delay elements, whose results are shown in Fig. 6(c) and (d), exhibit much lower levels of distortion, both in terms of insertion loss and group delay. However, this improvement comes at the cost of a substantial increase in the low-frequency insertion loss and a slight increase in the rate at which it rolls off at high frequency. While the compensation scheme addresses the distortion introduced by the delay elements, it has significant implications on how the tap multipliers are designed. Section 2.2.2 will address this matter in greater detail.

2.2.2. *Active elements: DACs, multipliers, and adders*

The parameters of the 7-tap feedforward equalizer are set electronically by writing values to a bank of seven registers. These in turn are connected to seven differential current mode DACs via a 98 wire bus, 49 wires for the non-inverted signals and 49 for the inverted signals. The DAC, which is shown in Fig. 7, consists of a set of binary-weighted current sources CS_{700} to CS_{706} and steering switches M_{700} to M_{713}. By combining the outputs of the current sources, any current value I_{dac} from $I_{ref}/8$ to $I_{ref} \times 8$ may be formed along with its complement. In order to ensure a constant voltage drop across the drain and source terminals of the steering switches, their gate widths must follow the same binary-weight scaling.

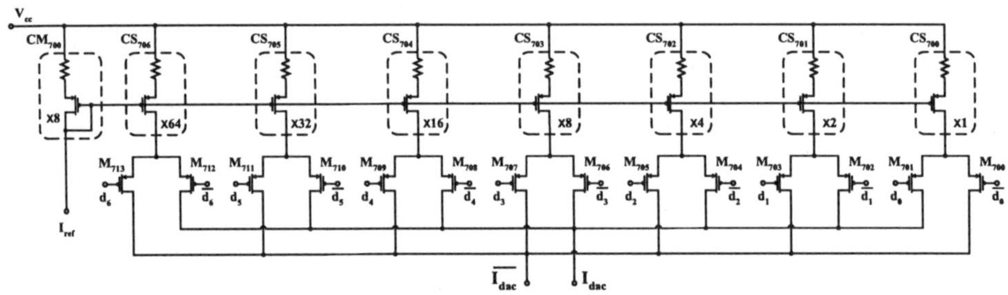

Fig. 7. 7-bit digital to analog converter.

The currents from the DACs, denoted as I_{dac} and its complement $\overline{I_{dac}}$, form the tap weight inputs to the FFE's four-quadrant multipliers. This variable gain block is based on the translinear gain cell[17-19]. Fundamentally, a translinear gain cell acts as a highly linear current scaler. The analog multiplier is merely a cross-coupling of two such gain cells. The present implementation of the translinear multiplier is divided into two separate sections, shown in Fig. 8 and Fig. 9. The first section, which is referred to as a linear to logarithmic converter takes as its input the DAC output currents and converts them to a differential voltage by means of identical base-emitter diodes Q_{802} and Q_{803}. By selecting R_{801}/R_{802} such that their values are equal to the sum of the emitter resistances of Q_{800}/Q_{801} and the resistors R_{803}/R_{804}, the differential input voltage is translated directly to the output at V_{mult} and $\overline{V_{mult}}$. These levels are then applied to the second section, shown in Fig. 9, which consists of two cross-coupled differential gain cells defined by Q_{903}/Q_{904} and Q_{905}/Q_{906}, respectively. If the base-emitter junction of the diodes of Q_{802} and Q_{803} are matched to the base-emitter junctions of the transistors Q_{903}/Q_{904} and Q_{905}/Q_{906}, then the currents at DOUTP and DOUTN will be proportional to the DAC input currents.

The multiplicative action in the circuit occurs because the differential input voltage signal at DINP and DINN controls the division of current between the cross-coupled differential gain cells. By inserting emitter resistances R_{903}, R_{904} and R_{905} having a sum that is much greater than the emitter resistance of Q_{901} and Q_{902}, the division of the current will be defined by the ratio of their sum to the sum of the collector resistors R_{901} and R_{902}. Being inherently linear circuit elements whose values do not vary with the input signal level, the current division defined by the resistors will be proportional to the amplitude of the differential input signal at DINP and DINN. In this way, the multiplier's differential output voltage must be a linear function of the product of both the DAC input

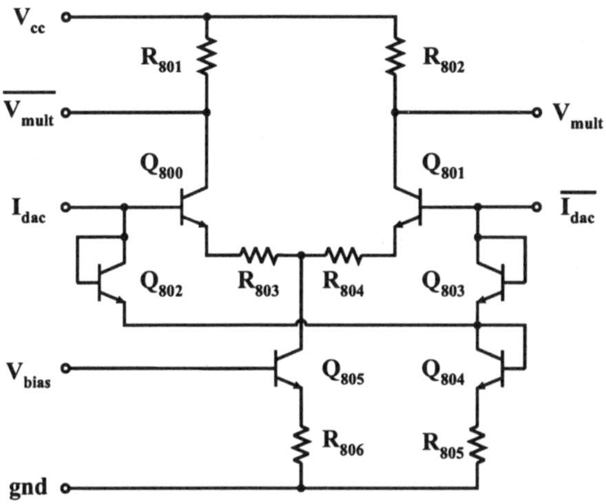

Fig. 8. Linear input to logarithmic output converter.

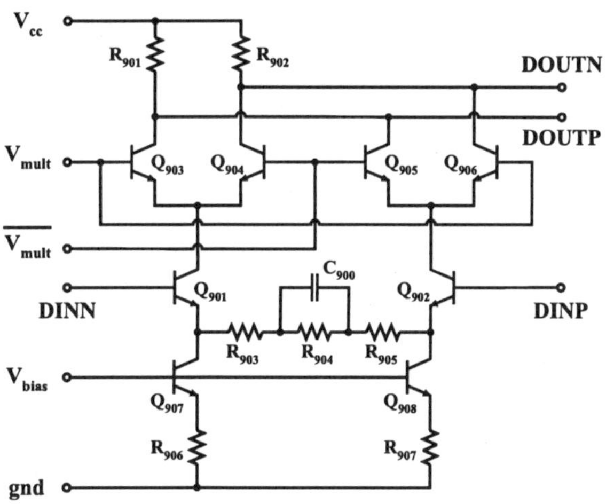

Fig. 9. Four-quadrant multiplier with bandwidth compensation elements.

and the data input. Since both the differential inputs and outputs can assume positive and negative polarities, the circuit is also referred to as a four-quadrant multiplier.

The translinear multiplier thus described is a well-known circuit topology that is used extensively in communication systems. Although the circuit can have a wide bandwidth with respect to the data rate and can readily accommodate the fixed insertion loss of the delay elements (by scaling the values of the load resistors R_{901} and R_{902} for each multiplier), the conventional design does not provide any compensation for the input bandwidth reduction and group delay variation. The absence of this feature is addressed in the present design by introducing a peaking capacitor C_{900} in parallel with R_{904}. Since the multiplier's output level is proportional to the ratio of the sum of the collector and emitter resistances, the capacitor will make the emitter resistance appear smaller as the frequency increases, thus reducing the amount of degeneration and increasing the gain.

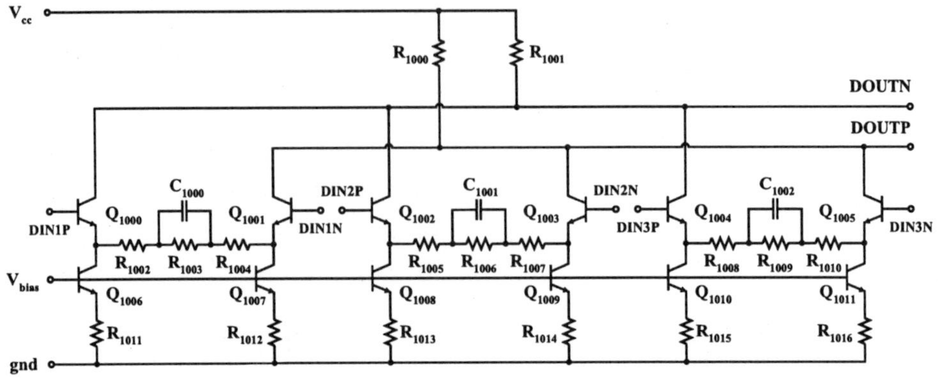

Fig. 10. Three input adder with bandwidth compensation elements.

By judicious selection of the values of C_{900} and R_{904}, the multiplier's complex gain-frequency profile can be made to compensate for some or all of the impairments introduced by the delay elements. If C_{900} and R_{904} are selected in the same way for each multiplier in the FFE, the outputs of each one can be made to have a nearly constant gain and group delay response with respect to the overall FFE input in the bandwidth of interest.

The adders, which appear as shown in Fig. 10, have three differential inputs and one differential output. Provided that none of the input transistors Q_{1000} to Q_{1005} are fully switched, each differential pair will behave like a fully independent linear amplifier with its voltage gain defined by the ratio of the sum of the collector resistances to the sum of its own emitter resistances. Likewise, the gain-profile for each one can be modified by adjusting its emitter resistor ($R_{1003}/R_{1006}/R_{1009}$) and capacitor values ($C_{1000}/C_{1001}/C_{1002}$). Introducing compensation into both adders and multipliers provides enhanced flexibility when fine-tuning the overall performance of the FFE circuit. By distributing the compensation elements among more blocks in the circuit, it also helps to keep their parameter values within realizable levels.

The complete addition function has been sub-divided into two hierarchical stages. In so doing, the capacitance loading the final output is reduced and its pole increased. Having an additional adder stage also allows for further distribution of the bandwidth compensation.

2.2.3. *FFE performance*

The effectiveness of the FFE can, in large part, be described by the static performance of the transversal filter. To this end, *S*-parameter measurements were made with each tap turned on to +1 and the AGC set to 0 dB. When not turned on, the remaining six taps were all set to their lowest value of +0.008. Since the VNA used in this test is a 2-port instrument, only single-ended measurements were possible. In other words, port 1 was AC coupled to the non-inverting input and port 2 was AC coupled to the non-inverting and un-retimed output (EQU OUT in Fig. 1). The inverting ports were both AC

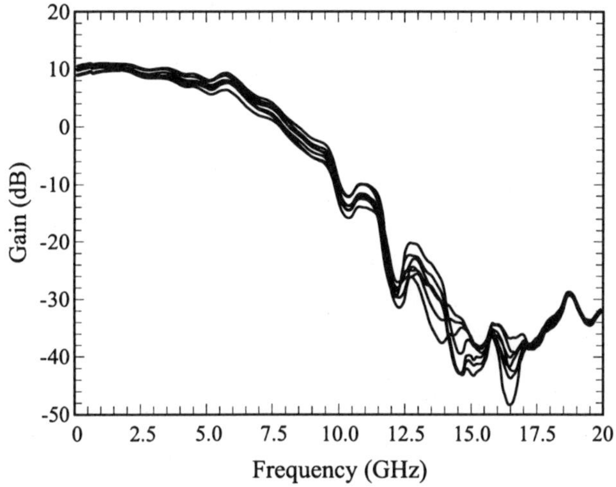

Fig. 11. Transversal filter single-ended insertion loss for each tap set to +1.

terminated to 50 Ω. To accommodate the fact that half the input power is dissipated in the inverting output load resistor, the S_{21} values are all shifted upwards by 6 dB.

The resulting measurements are shown in Fig. 11. Over the seven taps of the equalizer, the bandwidth is approximately 6.5 GHz, although it should be noted that this figure includes the effect of the AGC and the circuitry that routes the signal from the equalizer output to the EQU OUT test port. Consequently, it can be assumed that the equalizer performance is somewhat better than the measurement suggests. Comparing the gains of the different taps, there appears to be a 1.7 dB variation from tap 0 to tap 6. This is because the transmission line test structure had a slightly lower resistivity than the actual metal on the circuit. In both cases, the resistivity was within the limits specified by the IC process. Although significant, a 20% variation between first and last tap output will not have a dramatic impact on the overall equalizer performance.

In order to assess the linearity of the DAC outputs with respect to the tap gains, Fig. 12 presents a complete sweep of every FFE gain setting for a 100 MHz sine-wave input. The results are normalized to the highest output amplitude, namely tap 0 in the inverting state. Over all possible states, the measurement deviates from an ideal linear response by no more than 10%.

Fig. 13 and Table 1 illustrate the performance of the tap pathways in the time domain. In the first instance, Fig. 13 shows the phase response of the FFE with a 160 MHz square wave input. The relative tap delay is 36 ps with a peak variation of ±0.1 ps. Table 1 shows the relative tap performance using a 2^{31}-1 pseudorandom bit sequence (PRBS). The AGC is set to 0 dB gain and the input amplitude is kept to 100 mV_{pp}/side. This ensures that there is no clipping at any point in the linear path from the input of the IC to the EQU OUT port. Table 1 compares the tap performance in terms of three parameters, pathway Q, peak-to-peak jitter J_{pp}, and root-mean-squared (rms) jitter J_{rms}. The first of these is derived from the measurements as follows

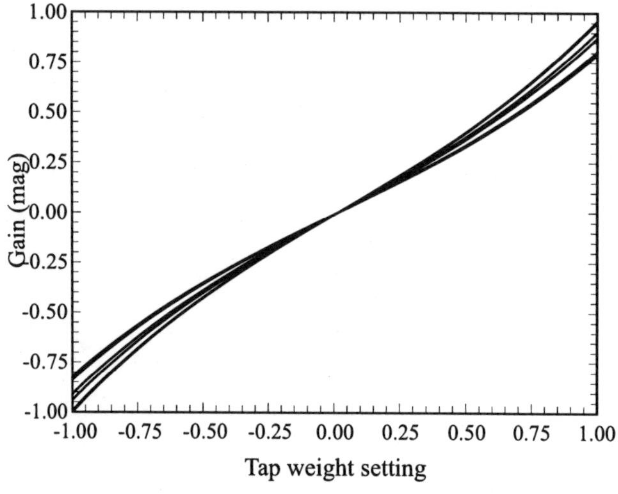

Fig. 12. Transversal filter low frequency gain versus tap and tap weight setting.

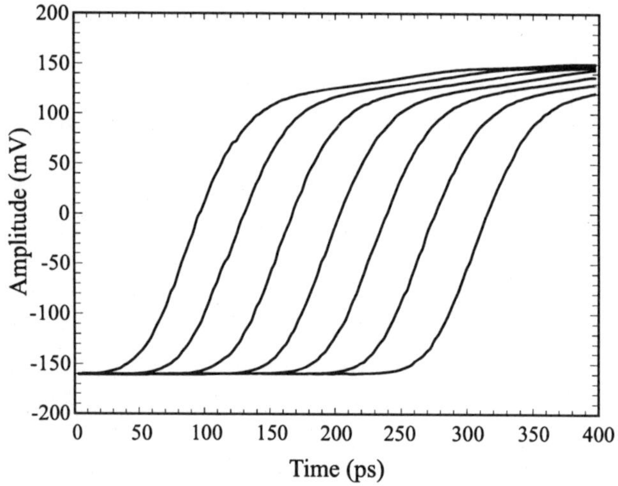

Fig. 13. Measured tap delay for a 160 MHz square-wave input.

$$Q[\text{dB}] = 20 \times \log\left(\frac{v_{pp}}{v_{rms,0} + v_{rms,1}}\right) \tag{12}$$

where v_{pp} is the peak-to-peak amplitude, $v_{rms,0}$ is the rms noise on the *zero*, and $v_{rms,1}$ is the rms noise on the *one*. Strictly speaking, the latter two terms are not actually noise because most of the spreading in the zero and one values is due to bandwidth limitation, consequently it is pattern dependent. Nonetheless, the Q value provides a good overall comparative figure of merit for the tap pathways.

From the table, it is readily apparent that taps 3-5 are slightly worse in performance than taps 0-2 and 6. This arises because the taps closest to the end points are terminated to the delay line's characteristic impedance, which gives minimal reflections. Towards

Table 1. Tap pathway performance metrics with a 2^{31}-1 PRBS pattern.

	Tap 0	Tap 1	Tap 2	Tap 3	Tap 4	Tap 5	Tap 6
Q [dB]	16.4	16.2	15.7	15.0	14.6	14.6	16.0
J_{pp} [ps]	26.7	28.5	28.5	32.8	40.5	36.2	33.2
J_{rms} [ps]	5.2	5.8	5.9	6.7	7.6	8.0	6.3

the center taps and farthest from the terminations, the impedance deviates from the matched condition owing to skin-effect losses and the deliberate insertion of the shunt resistors. This necessarily implies that, if the delay line is broken at the centre tap, the impedance seen looking towards the load and towards the input will have identical real and imaginary components. Although this might appear to be a propitious condition to have, it actually leads to a significant mismatch. For a perfect match at the center tap, the impedance seen looking towards the input should be the *conjugate* of the impedance seen looking towards the load. Because of this mismatch, all of the metrics, including the jitter tend to be worst at the centre taps. The slight shift in poorer performance towards tap 4 over tap 3 is because of slight asymmetries in the layout and in the source and load terminations.

2.3. *Decision feedback equalizer*

The function of the DFE is to remove the effect of postcursor ISI. This is achieved by subtracting a portion of the previously decided symbol with the current symbol before its value is decided. For the case of a single tap DFE, the current bit under decision is reduced by a certain fraction if the previously decided bit was a one and increased if it was a zero. In the event that the current bit is a zero and the previous bit was a one, the reduction will strengthen the zero at a modest cost to the strength of the one. The fraction referred to previously is simply the DFE tap weight.

A top-level block diagram of the DFE is presented in Fig. 14. It consists of an adder, a comparator circuit, a latch, and a multiplier. Although the present design consists of just one single tap, this architecture can be readily extended to multiple taps by adding additional latches and multipliers. The design of the DFE loop presents a significant

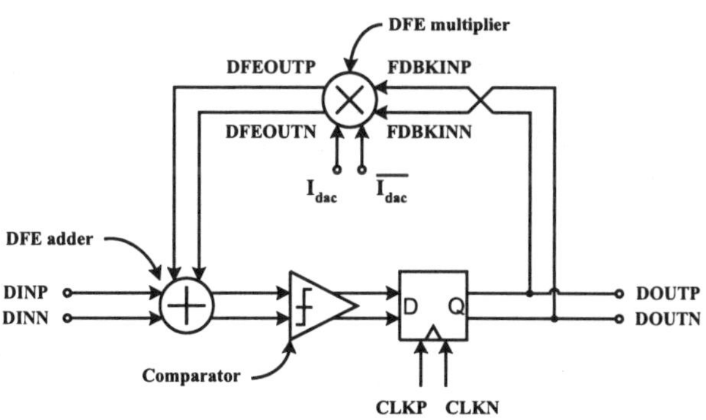

Fig. 14. Block diagram of the decision feedback equalizer.

challenge because it must operate at the full-bit rate. This requires particular attention to the loop propagation time. If it deviates significantly from one unit interval (UI), the equalizer's performance is significantly compromised and the loop may become unstable. Notwithstanding this requirement, having a CDR integrated directly with the equalizer provides an added measure of control because it ensures that the re-timing latch is always triggered at the center of the eye. As a final note, the function of the DFE is to subtract a fraction of the previous symbol. In the present implementation, which includes an adder, the subtraction operation is realized by inserting a cross-over in the feedback path just before the tap multiplier.

2.3.1. *Active elements: adder and multiplier*

The DFE adder operates according to the same principles as those described in relation to Fig. 10. As in the FFE, the adder takes as its input a current, which is derived from the DFE multiplier. For its part, the DFE multiplier circuit is shown in Fig. 15 and is described as follows. Fundamentally, the differential pair defined by Q_{1500} and Q_{1501} switch current between DFEOUTP and DFEOUTN at the full bit rate in response to the polarity of the retimed symbols received by the FDBKINP and FDBKINN inputs. For example, if the previous bit was a one, FDBKINP will be high, which in turn implies that the DFE control circuit will sink current through the DFEOUTN terminal. As a result, the inverted terminal of the DFE adder will be lowered, thus favouring the detection of a zero. The magnitude of the switched current I_{1500} is controlled by the current source defined by Q_{1506} and its degeneration resistor R_{1501}. It in turn derives its value from the current mirror Q_{1509} and R_{1500}, the input to which I_{dac} is generated from a DAC identical to the one depicted in Fig. 7. Q_{1504} is a level shifting diode that ensures that Q_{1506} does not

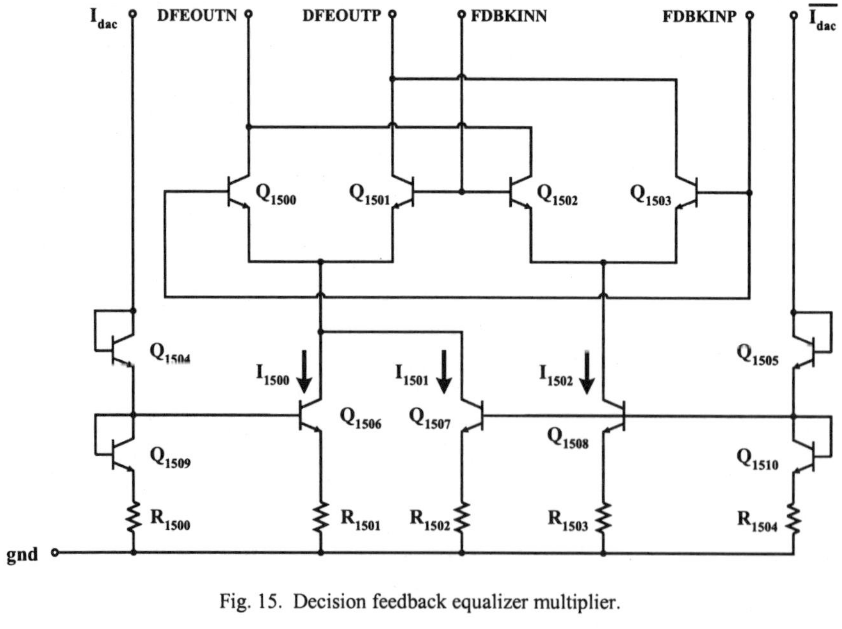

Fig. 15. Decision feedback equalizer multiplier.

become saturated.

Since the differential pair Q_{1500} and Q_{1501} are switching at 10 Gb/s, it is imperative that the source current remain constant regardless of the applied DFE tap weight value. For this reason, a second current source, Q_{1507} and R_{1502} is connected in parallel with I_{1500}. Its value, I_{1501}, is derived from the DAC in the same way as I_{1500}, only it takes as its input the complementary DAC current $\overline{I_{dac}}$. Consequently, when I_{1500} is reduced for low DFE tap values, it will increase so as to maintain the same total switching current. The last portion of the circuit, the differential pair Q_{1502} and Q_{1503} and current source Q_{1508} and R_{1503} is included simply to cancel the impact of the compensatory current I_{1501}. This is possible because the second differential pair is connected with the opposite polarity to Q_{1500} and Q_{1501}. Since I_{1502} has the same value as I_{1501}, being derived from the same current mirror, all impact of I_{1501} on the main differential pair is removed. To re-iterate, this scheme allows the main differential DFE control pair to operate with a constant switching current, while still permitting the feedback level to be varied.

2.3.2. *DFE performance*

By definition, a decision feedback equalizer can only mitigate post-cursor ISI. To test the DFE's performance, an electrical post-cursor heavy "channel" was created with a set of Bessel-Thompson filters having an overall 3 dB bandwidth of approximately 4 GHz. When combined with the requisite attenuators and test cables to adjust the level and route the 10 Gb/s signal to the input of the IC, the bandwidth becomes smaller still. Using a bit error rate tester (BERT) to generate a 10 Gb/s signal with a 2^{31}-1 pattern, it was observed that the isolated ones and zeros in the sequence had an amplitude of 10 mV_{pp}/side for a mean input signal level of the order of 100 mV_{pp}/side. The DFE works well with this sort of bandwidth impaired signal because it can boost the amplitude of the isolated ones and zeros, without significantly degrading the amplitude of the long consecutive runs.

For the first measurement, a 10 Gb/s signal was passed through the channel, recovered by the IC with the AGC disabled, and routed back to the error detector. Although not shown in the top level block diagram of Fig. 1, the equalizer/CDR IC has a mode whereby the FFE can be bypassed, thus allowing for an independent measurement of the DFE. The DFE tap weight was then swept from its lowest value to its highest value and the bit error rate (BER) measured for each value. The results, which are shown in Fig. 16, are quite dramatic. In the first instance, they show that even for a simple channel, the CDR can do no better than a BER of 10^{-2} without DFE. With the DFE running, it can make a significant difference as in the case of Fig. 16(a). This shows the results for a 50 mV_{pp}/side input signal with an eye opening of approximately 5 mV_{pp}/side. By optimizing the tap weight setting, it is possible to reduce the BER by over 10 orders of magnitude. Increasing the DFE beyond the optimum has a steadily deteriorating effect. Although the isolated ones and zeros are now quite large, the effect of too much feedback is now lowering the amplitude of the consecutive runs such that more errors are being generated than removed. Further increases to the DFE tap setting eventually leads to error multiplication, although it should be noted that the circuit still remains inherently stable.

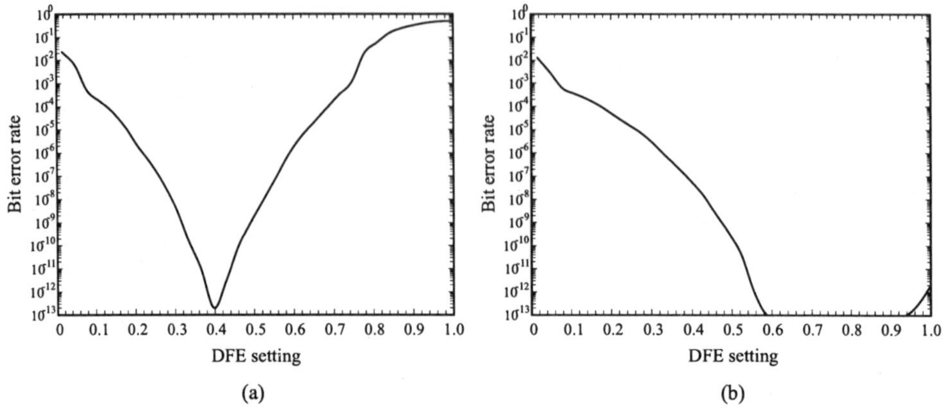

Fig. 16. Decision feedback equalizer BER performance versus tap weight setting: (a) 50 mV_{pp}/side input, and (b) 100 mV_{pp}/side input.

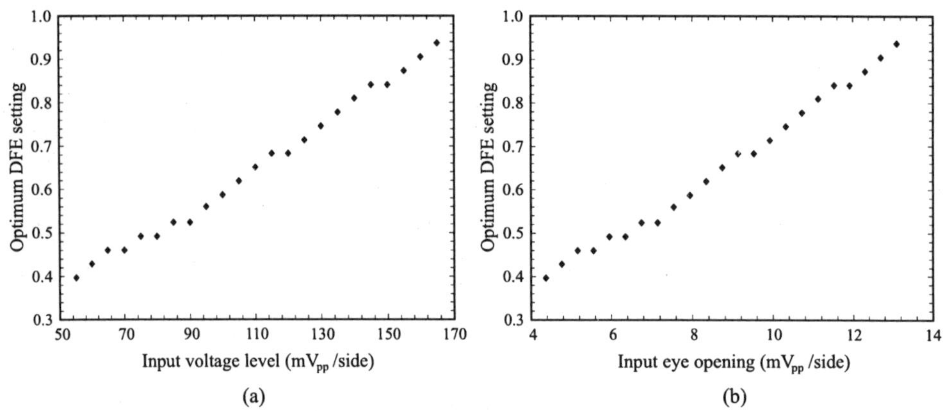

Fig. 17. Decision feedback equalizer setting for BER < 10^{-12} versus: (a) input signal amplitude, and (b) minimum input eye opening.

Fig. 16(b) shows the same measurement with a larger amplitude of 100 mV_{pp}/side. Not only does the error floor drop to well below 10^{-13}, but the optimum point shifts upwards by a factor of two, which is consistent with the doubling in input amplitude.

For the second measurement, the input signal amplitude was swept over a range of 50 mV_{pp}/side to 170 mV_{pp}/side with the same channel. For each amplitude setting, the DFE tap weight was optimized to achieve a BER of less than 10^{-12}. The results, shown in Fig. 17(a) versus input eye amplitude and Fig. 17(b) versus minimum input eye opening, indicate a strong linear dependence on DFE setting versus input signal level. This is consistent with the previous measurement and also indicates the linearity of the applied tap weight setting to the actual signal level fed back. In other words, it validates the design of the DFE multiplier depicted in Fig. 15.

2.4. *Clock and data recovery*

The CDR is based on a charge-pump phase-locked loop (PLL) architecture[25], of which the key blocks in the design are shown in Fig. 18. They include a linear phase detector that compares the relative position of the clock edges to the data edges, a charge pump that controls the level on a loop filter and a voltage controlled oscillator (VCO). All circuits in the PLL are fully differential. In addition to retiming the data from the equalizer, the CDR also provides a clock signal to the latch in the DFE. The phase detector is designed to output a charge up or charge down pulse on every data transition, where the width of the pulse is a function of the degree of timing misalignment. These pulses are then used by the charge pump to raise or lower the level on an external RC loop filter, which in turn shifts the phase of the clock by means of the differential VCO. Although not shown, the circuit contains refinements that address frequency acquisition, phase detector gain variation with transition density, as well as a means to hold the clock in the center of the eye during long runs of consecutive bits. The CDR can operate with input data signals having bitrates of 9.95-11.1 Gb/s.

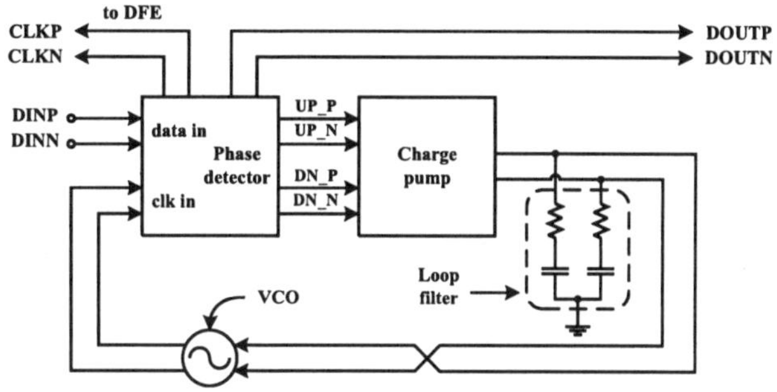

Fig. 18. Block diagram of the clock and data recovery.

3. Fabrication and packaging

The IC has dimensions of 3 × 3 mm^2 and was fabricated using a commercially available 0.18 μm SiGe BiCMOS technology. It is packaged in a 7 × 7 mm^2 plastic ball-grid array (PBGA).

4. Measurement results

The experimental setup for testing the equalizer is shown in Fig. 19. It consists of two actual equalizer ICs, one to emulate a dispersive channel and another, which is the device under test, to compensate for it. The channel is created by passing the input signal through the transversal filter of the first equalizer and setting multiple taps to non-zero values. The setup also includes a 10 Gb/s SERDES configured in loop-back mode, the purpose of which is to provide a phase stable output bit stream for the BERT. This is possible because the SERDES has a pair of first-in first-out (FIFO) memories. The principal advantage of this setup is that it allows a number of different channels to be

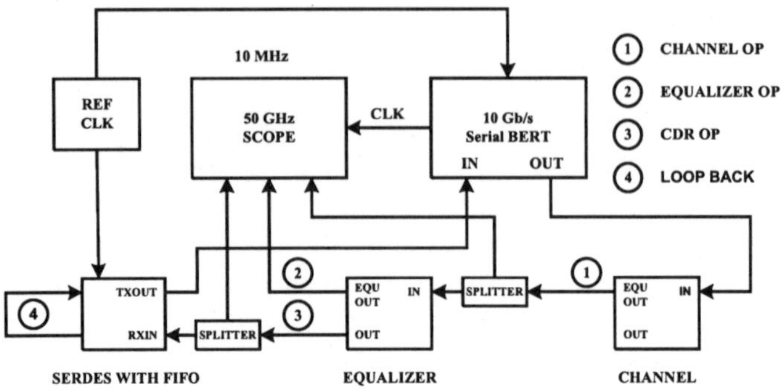

Fig. 19. Equalizer measurement test setup.

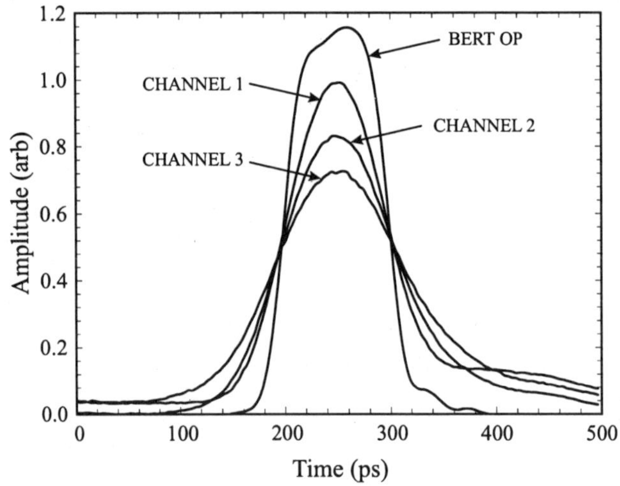

Fig. 20. Measured impulse response of three electrical channels with BERT output shown as a reference.

tested by the equalizer without requiring any reconfiguration of the hardware. However, since the "channel" is itself an active device, it introduces additional impairments to the input signal that must be taken into account. For example, the channel introduces noise and some slight nonlinear distortion, both of which have an adverse effect on the measured sensitivity.

Fig. 20 shows the impulse responses for three electrical channels with the BERT output given as a reference. Comparing the widths of the pulses at an amplitude of 0.2, these range from 160 to 220 ps for the channels and 120 ps for the undistorted BERT output. In terms of the eye diagram, only the first channel gives any eye opening, and it is of the order of 10 mV_{pp}/side for an input level of 70 mV_{pp}/side or -19 dBm/side. Channels 2 and 3 are much worse and exhibit no eye opening at all. All measurements are made at a bit rate of 10 Gb/s with a 2^{31}-1 PRBS pattern.

As an illustration of the equalizer's performance, Fig. 21 shows the eye diagrams for channel 3 at points 1, 2 and 3 of Fig. 19. As noted previously, the input, Fig. 21(a), shows no eye opening for a nominal input level of -10 dBm/side. After transiting the equalizer,

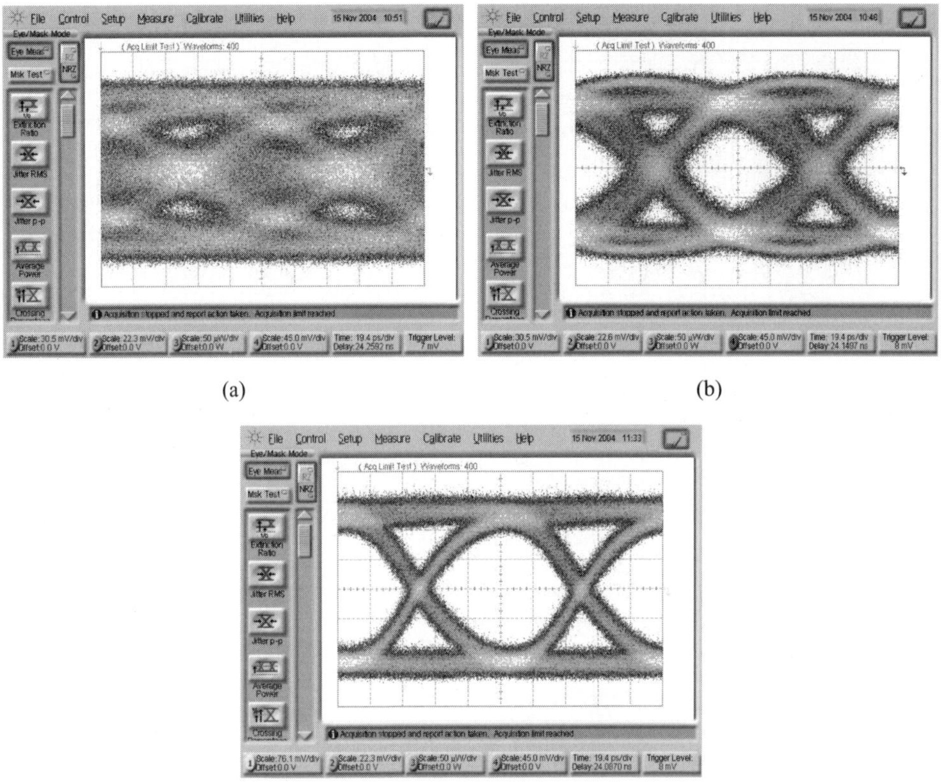

(a) (b)

(c)

Fig. 21. Eye diagrams showing performance for channel 3: (a) input after channel, (b) after equalizer at EQU OUT, and (c) after retiming by the CDR.

which includes corrections from both the FFE and DFE, the eye diagram is opened to the form shown in Fig. 21(b). The equalizer's tap weight parameters have been adapted by minimizing the measured BER. Following the equalizer, the signal is readily acquired by the CDR and retimed as shown in Fig. 21(c). The BER for these eye diagrams is approximately 10^{-7} and reflects the fact that there is some noise speckling inside the eye.

A more comprehensive set of measurement results is presented in Fig. 22, which shows the sensitivity curves for the three electrical channels. They depict BER versus electrical input signal level expressed in dBm/side. As a point of reference, the CDR exhibits a 10^{-12} sensitivity level below 10 mVpp/side or -36 dBm with no distortion at the input. Although seemingly benign, channel 1 still imposes a significant penalty on the equalizer's sensitivity, of the order of 16 dB. It should be noted, however, that a major contributor to this penalty is in the form of channel noise, rather than dispersion. Channels 2 and 3 exhibit additional penalties of 8 and 20 dB, respectively. In terms of ISI penalty reduction, the input signal level for channel 1 must be increased by 6.2 dB in order to obtain a BER of 10^{-12} without the equalizer. With respect to channels 2 and 3, the

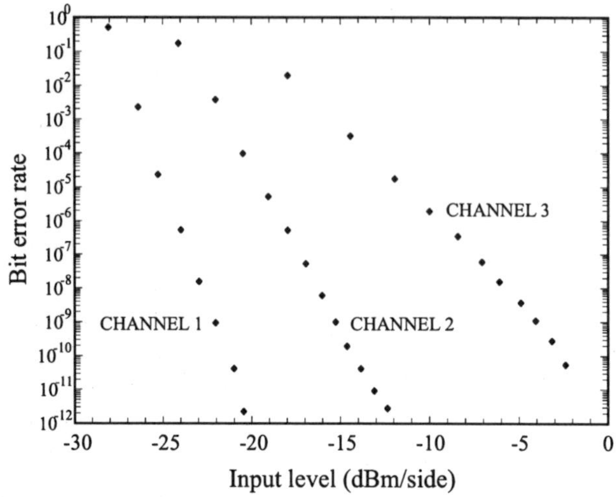

Fig. 22. Measured bit error rate curves of the equalizer for three electrical channels with a 2^{31}-1 PRBS pattern.

total absence of an eye opening means that there is no way to lock a CDR to the distorted signal. Hence, it is not possible to comment on ISI penalty reduction in these cases.

5. Conclusion

An architecture for a 10 Gb/s analog continuous time equalizer with integrated CDR has been designed and fabricating using a 0.18 μm BiCMOS technology. In terms of performance and cost, it has been asserted that the equalizer should be integrated directly with the CDR, and indeed, this is the form of the present architecture. Not only does it simplify the receiver design, but it improves signal integrity before the decision circuit and makes a synchronous clock available to the equalizer itself. For their part, the equalizer blocks have been extensively reviewed and characterized, the purpose being to highlight the importance of minimizing the circuit's own self-induced distortion. This requires careful mitigation of group delay variation and bandwidth erosion in the FFE and the application of innovative circuit design techniques for implementing the DFE. Although not discussed in any great detail, these statements apply equally to the AGC and CDR. By taking into consideration these important implementation issues, the resulting equalizer is able to recover signal degraded by significant amounts of dispersion, so much so that a conventional CDR would be rendered inoperable. For more benign signals, the IC shows an ISI penalty reduction of 6.2 dB for a fairly noisy channel.

References

1. K. Azadet, E. F. Haratsch, H. Kim, F. Saibi, J. H. Saunders, M. Shaffer, L. Song, and M.-L. Yu, "Equalization and FEC techniques for optical transceivers," *IEEE J. Solid-State Circuits*, **37**(3), 317-327 (2002).
2. M. D. Feuer, S.-Y. Huang, S. L. Woodward, O. Coskun, and M. Boroditsky, "Electronic dispersion compensation for a 10-Gb/s link using a direct modulated laser," *IEEE Photon. Tech. Lett.*, **15**(12), 1788-1790 (2003).

3. P. Pepeljugoski, J. Schaub, J. Tierno, J. Kash, S. Gowda, B. Wilson, and A. Hajimiri, "Improved performance of 10 Gb/s multimode fiber optic links using equalization," in *OFC 2003 Postconference Technical Digest*, Atlanta, GA, Mar. 2003, vol. 2, 472-474.
4. G. S. Kanter, A. K. Samal, and A. Gandhi, "Electronic dispersion compensation for extended reach," in *OFC 2004 Postconference Technical Digest*, Los Angeles, CA, Feb. 2004, TuG1.
5. M. Nakamura, H. Nosaka, M. Ida, K. Kurishima, and M. Tokumitsu, "Electrical PMD equalizer ICs for a 40-Gbit/s transmission," in *OFC 2004 Postconference Technical Digest*, Los Angeles, CA, Feb. 2004, TuG4.
6. H. F. Haunstein, W. Sauer-Greff, A. Dittrich, K. Sticht, and R. Urbansky, "Principles for electronic equalization of polarization-mode dispersion," *J. of Lightwave Technol.*, **22**(4), 1169-1182 (2004).
7. H. Bulow, and G. Thielecke, "Electronic PMD mitigation - from linear equalization to maximum-likelihood detection," in *OFC 2001 Postconference Technical Digest*, Anaheim, CA, Mar. 2001, WAA3.
8. J. Lee, P. Roux, U.-V. Koc, T. Link, Y. Baeyens, and Y.-K. Chen, "A 5-b 10-GSample/s A/D converter for 10-Gb/s optical receivers," *IEEE J. Solid-State Circuits*, **39**(10), 1671-1679 (2004).
9. W. Ellersick, C.-K. K. Yang, M. Horowitz, and W. Dally, "GAD: a 12-GS/s CMOS 4-bit A/D converter for an equalized multi-level link," in *VLSI Circuit Technical Digest of Papers*, Kyoto, Japan, Jun. 1999, pp. 49-52.
10. S. L. Woodward, S.-Y. Huang, M. D. Feuer, and M. Boroditsky, "Demonstration of an electronic dispersion compensator in a 100-km 10-Gb/s ring network," *IEEE Photon. Tech. Lett.*, **15**(6), 867-869 (2003).
11. J. Lee, A. Leven, J. S. Weiner, Y. Baeyens, Y. Yang, W.-J. Sung, J. Frackoviak, R. F. Kopf, and Y. K. Chen, "A 6-b 12-GSamples/s track-and-hold amplifier in InP DHBT technology," *IEEE J. Solid-State Circuits*, **38**(9), 1533-1539 (2003).
12. X. Wang and R. R. Spencer, "A low-power 170-MHz discrete-time analog FIR filter," *IEEE J. Solid-State Circuits*, **33**(3), 417-426 (1998).
13. P. Popescu, H. Tran, D. S. McPherson, "High speed circuits for electronic dispersion compensation," U.S. patent application Ser. No. 10/638,386, filed Aug. 12, 2003.
14. P. Popescu, D. Gradinaru, K. Howlett, H. Tran, "A differential receiver circuit with electronic dispersion compensation," U.S. patent application Ser. No. 10/638,387, filed Aug. 12, 2003.
15. P. Popescu, D. S. McPherson, S. Szilagyi, H. Tran, and K. Howlett, "A differential receiver circuit with electronic dispersion compensation for optical communication systems," U.S. patent application Ser. No. 10/638,340, filed Aug. 12, 2003.
16. P. Popescu, "Quake Technologies Inc. electronic dispersion compensation," Quake internal document, Sep. 2003.
17. B. Gilbert, "A new wide-band amplifier technique," *IEEE J. Solid-State Circuits*, **3**(4), 353-365 (1968).
18. B. Gilbert, "A precise four-quadrant multiplier with subnanosecond response," *IEEE J. Solid-State Circuits*, **3**(4), 365-373 (1968).
19. H.-M. Rein, L. Schmidt, K. Wörner, and W. Pieper, "Wide-band symmetrical analog multiplier IC for coherent optical-fiber receivers operating up to 10 Gb/s," *IEEE J. Solid-State Circuits*, **26**(12), 1840-1846 (1991).
20. H. Wu, J. A. Tierno, P. Pepeljugoski, J. Schaub, S. Gowda, J. A. Kash, and A. Hajimiri, "Integrated transversal equalizers in high-speed fiber-optic systems," *IEEE J. Solid-State Circuits*, **38**(12), 2131-2137 (2003).
21. J. Lee and A.P. Freundorfer, "MMIC adaptive transversal filtering using Gilbert cells and is suitable for high-speed lightwave systems," *IEEE Photon. Tech. Lett.*, **12**(2), 196-198 (2000).

22. W. R. Eisenstadt, Y. Eo, "*S*-parameter-based IC interconnect transmission line characterization," *IEEE Trans. Comp., Hybrids, and Manufact. Technol.*, **15**(4), 483-490 (1992).
23. R. A. Pucel, D. J. Masse, and C. P. Hartwig, "Losses in microstrip," *IEEE Trans. Microwave Theory Tech.*, **16**(6), 342-350 (1968).
24. R. E. Collin, *Foundations for microwave engineering: 2nd edition* (McGraw-Hill, New York, 1992).
25. D. A. Johns and K. Martin, *Analog integrated circuit design* (John Wiley & Sons, New York, 1997).

International Journal of High Speed Electronics and Systems
Vol. 15, No. 3 (2005) 549–566

World Scientific
www.worldscientific.com

Equalizer Architectures for 40-Gb/s Optical Systems Limited by Polarization-Mode Dispersion

Jonathan Sewter
Anthony Chan Carusone[§]
Edward S. Rogers Sr. Department of Electrical and Computer Engineering
University of Toronto
10 King's College Rd.
Toronto, Canada, M5S 3G4
[§] *tcc@eecg.utoronto.ca*

An analysis of first-order polarization-mode dispersion (PMD) effects in a 40-Gb/s optical system is used to compare different electronic equalizer architectures as potential PMD compensators. Both linear and nonlinear equalizers are considered employing symbol-spaced and fractionally-spaced taps. It is found that a decision feedback equalizer consisting of a 3-tap symbol-spaced feedforward equalizer and a 1-tap feedback equalizer effectively eliminates PMD as the dominant length-limiting factor in most 40-Gb/s optical systems. Such an equalizer would entail less complexity than several previously reported electronic PMD compensators.

Keywords: optical communication, polarization mode dispersion, adaptive equalizer

1. Introduction

Optical communication systems have been used since the 1970s for high-volume data transmission within wide-area, metropolitan-area and local-area networks[1]. Until recently, long-haul links over single-mode fiber (SMF) could be designed without concern for the bandwidth limitations of the fiber. By compensating for fiber loss with amplifiers, the reach of these systems could be extended. To satiate the demand for greater network capacity, the data rate of current optical systems has been pushed to 10 and 40 Gb/s (OC-192 and OC-768). At these data rates, it is no longer possible to neglect the bandwidth limitations of SMF, as several dispersion mechanisms lead to frequency-dependent loss[2].

The two most important dispersion mechanisms for SMF are chromatic dispersion (CD) and polarization-mode dispersion (PMD). CD is a result of the wavelength-dependency of the refractive index of the fiber. PMD results from the variation in the refractive index of the fiber with respect to the polarization of the light signal. Since CD can be compensated by proper choice of optical fiber, PMD has been identified as the limiting factor in high-speed optical systems[3].

To mitigate the effects of PMD, optical systems must include some form of PMD compensation. This compensation can be achieved either optically or electronically.

Electronic PMD compensation schemes are attractive because they allow greater integration with existing circuitry, leading to more compact, less expensive solutions. This is especially true for wavelength-division multiplexed (WDM) systems, in which every channel needs PMD compensation[4]. Also, because PMD fluctuates with changes in temperature and environment, PMD compensators must be able to adapt to varying channel conditions within milliseconds[5]. Fast and accurate adaptation is more easily performed in the electronic domain. Successful electronic equalization has already been demonstrated at 10 Gb/s by Wu et al (7-tap FFE)[6], Bülow et al (8-tap FFE, 1-tap FBE)[7], and Möller et al (1-tap FBE)[8], and more recently at 40 Gb/s by Nakamura et al (3-tap FFE, 1-tap FBE)[9], and Hazneci and Voinigescu (7-tap FFE)[10].

While it has been shown that nonlinear equalization using a decision feedback equalizer (DFE) is required to reduce the power penalty caused by PMD to acceptable levels[11], the tradeoffs between different equalizer architectures are not evident. In this paper, a system-level analysis of PMD effects in a 40-Gb/s system using Matlab/Simulink is used to compare several equalizer architectures in terms of overall system performance, as has been done for optical compensation schemes[12]. Both linear and decision feedback equalizer (DFE) architectures are considered employing symbol-spaced and fractionally-spaced taps. Using this analysis we are able to quantify the increase in system reach that can be attained by equalization. Also, we show that effective compensation of PMD can be achieved using fewer taps in the feedforward equalizer (FFE) and feedback equalizer (FBE) than have been used in several previously reported implementations.

In Section 2 of this paper, the effects of PMD on optical transmission are reviewed and in Section 3 PMD compensation methods are broadly surveyed. Section 4 provides a description of the system model used for the Matlab/Simulink simulations. In Section 5 the various equalizer architectures considered for PMD compensation are described and Section 6 explains the methodology of the simulations performed. Finally, Section 7 contains the results of these simulations and Section 8 concludes the paper.

2. Polarization-Mode Dispersion (PMD)

PMD is a result of the phenomenon of birefringence which affects all real optical fibers. Birefringence refers to the difference in refractive index experienced by light in the two orthogonal polarization modes of the fiber. It is caused by ellipticity of the fiber cross-section due to asymmetric stresses applied to the fiber during or after manufacturing. Birefringence leads to fast and slow modes of propagation and consequently dispersion[13].

In terms of digital communications, PMD results to a first order in an input pulse being split into a fast and slow pulse which arrive at the receiver at different times, as shown in Figure 1. If the differential delay of the two pulses is significant compared to the bit period, intersymbol interference (ISI) and an increase in bit-

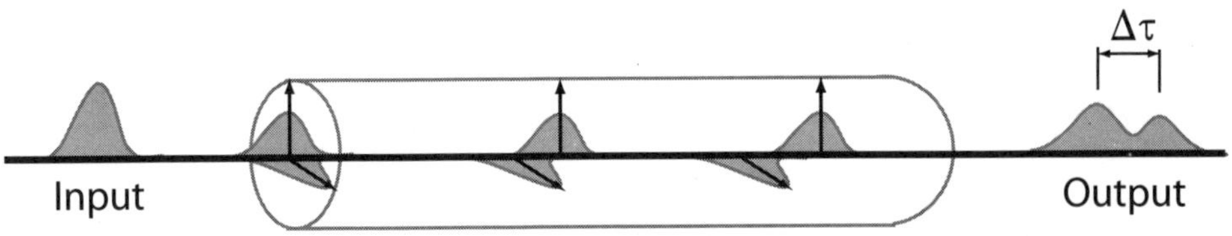

Fig. 1. Pulse bifurcation due to PMD. The power in the input pulse is split between the two polarization modes of the fiber. Birefringence causes a difference in phase velocities between the two modes, resulting in ISI at the output.

error rate (BER) will result.

2.1. *Consequences of PMD for Optical Systems*

To a first order, the impulse response of an optical fiber with PMD is[14]:

$$h_{\text{PMD}}(t) = \gamma\delta(t) + (1-\gamma)\delta(t-\Delta\tau) \tag{1}$$

where γ is the proportion of the optical power in the "fast" state of polarization (SOP), (1-γ) is the proportion of power in the "slow" SOP and $\Delta\tau$ is the differential group delay (DGD) between the fast and slow components.

The parameters γ and $\Delta\tau$ vary depending on the particular fiber and its associated stresses; γ can take any value from zero to one with uniform probability throughout this range[15], and $\Delta\tau$ varies statistically according to a Maxwellian distribution[16] given by:

$$\rho(\Delta\tau) = \sqrt{\frac{2}{\pi}}\frac{\Delta\tau^2}{\sigma^3}e^{-\frac{\Delta\tau^2}{2\sigma^2}} \tag{2}$$

The constant σ in (2) is related to the average DGD, $\Delta\tau_{avg}$, by[17]:

$$\sigma = \frac{\sqrt{2\pi}\Delta\tau_{avg}}{4} \tag{3}$$

Therefore, though it can occasionally assume large values, $\Delta\tau$ will for the most part remain close to its average value. Furthermore, $\Delta\tau$ varies over time with significant variations sometimes observed on the order of milliseconds[5].

The average DGD per unit length of a fiber is defined as its PMD parameter, which has units of ps/$\sqrt{\text{km}}$. Typical installed fibers exhibit a PMD of 0.5 - 2.0 ps/$\sqrt{\text{km}}$[18]. New fibers can be manufactured with a PMD of as low as 0.05 ps/$\sqrt{\text{km}}$[19]. Given the PMD parameter, the average DGD of a fiber of length L is given by:

$$\Delta\tau_{avg} = \text{PMD} \times \sqrt{L} \tag{4}$$

It has been calculated that to prevent PMD from causing system outages amounting to more than thirty seconds per year (corresponding to an outage probability of 10^{-6}), the average DGD must be less than approximately 15% of a bit period, T_B[3].

$$\Delta\tau_{avg} < 0.15T_B \tag{5}$$

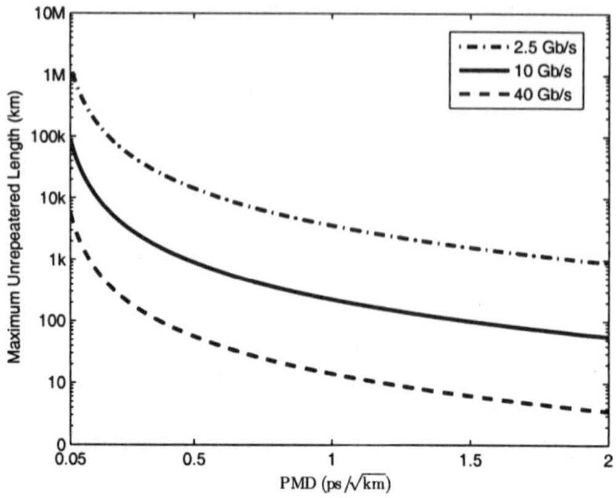

Fig. 2. Plot of maximum unrepeatered link length for fibers with varying PMD parameters for 2.5-, 10- and 40-Gb/s optical systems, assuming PMD is the dominant limiting factor.

This has severe implications as the data rate of these systems is increased to 10 and 40 Gb/s. As the data rate is increased on a given fiber, the maximum useful length of the fiber decreases according to the square of the increase. For example, given a fiber with a PMD of 1.0 ps/$\sqrt{\text{km}}$ and using (5), the maximum length of a 2.5-, 10- and 40-Gb/s system is 3600, 225 and 14 km, respectively, if PMD is the limiting factor. This relationship is illustrated in Figure 2.

2.2. *PMD Frequency-Domain Analysis*

The frequency-domain characteristic of an optical fiber with PMD can be easily obtained by taking the Fourier transform of its impulse response, $h_{\text{PMD}}(t)$, as defined in (1). The transfer function, $H_{\text{PMD}}(f)$, is described by:

$$H_{\text{PMD}}(f) = \gamma + (1-\gamma)e^{-j2\pi f \Delta\tau} \tag{6}$$

which is equal to:

$$H_{\text{PMD}}(f) = \gamma + (1-\gamma)[\cos(2\pi \frac{f}{f_{DGD}}) - j\sin(2\pi \frac{f}{f_{DGD}})] \tag{7}$$

where $f_{\text{DGD}} = \frac{1}{\Delta\tau}$. By inspection of (7), it can be seen that $|H_{\text{PMD}}(f)|$ has maximae at $f = kf_{\text{DGD}}, k \in I$, and minimae at $f = \frac{(2k-1)}{2} f_{\text{DGD}}, k \in I$.

Magnitude and phase plots of $H_{PMD}(f)$ for varying PMD conditions (γ and $\Delta\tau$) are shown in Figure 3.

From these plots it is apparent that the frequency response of a PMD fiber varies greatly depending on the specific nature of the PMD conditions. In general, PMD can cause notches in the frequency response of the fiber. The frequency of these notches is proportional to the DGD. The depth of these notches is dependent on γ, with the case $\gamma = 0.5$ resulting in nulls. The wide variation in potential frequency

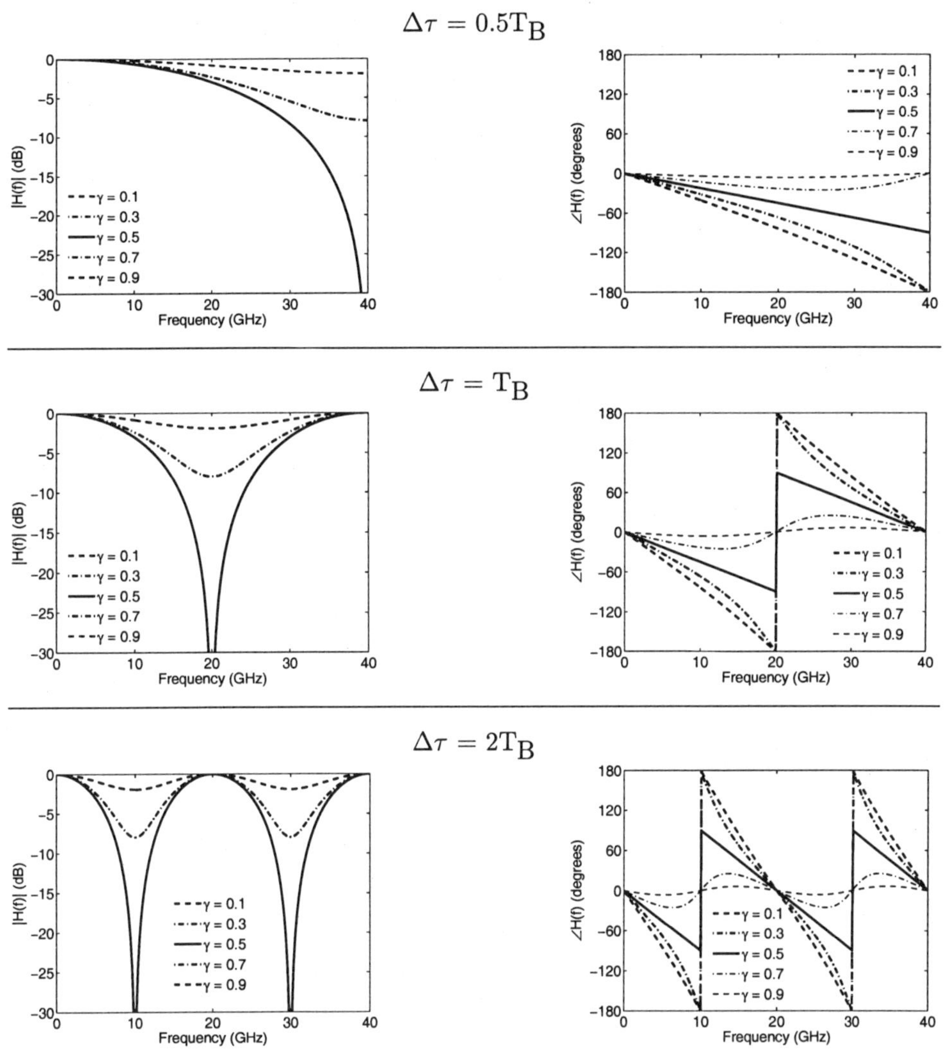

Fig. 3. Magnitude and phase responses of PMD channels with varying γ and $\Delta\tau$ characteristics. T_B is a bit period (e.g. $T_B = \frac{1}{40Gb/s} = 25$ ps at 40 Gb/s).

responses and the possibility of nulls in the frequency spectrum make equalization of this channel difficult.

3. PMD Compensation Methods

Forward error correction (FEC)[20,21] and wavelength redundancy in WDM networks[22] have been suggested as means of mitigating the effects of PMD. However, direct compensation of PMD effects is often required either independent of or in conjunction with redundancy schemes[23,24]. PMD can be directly compensated in

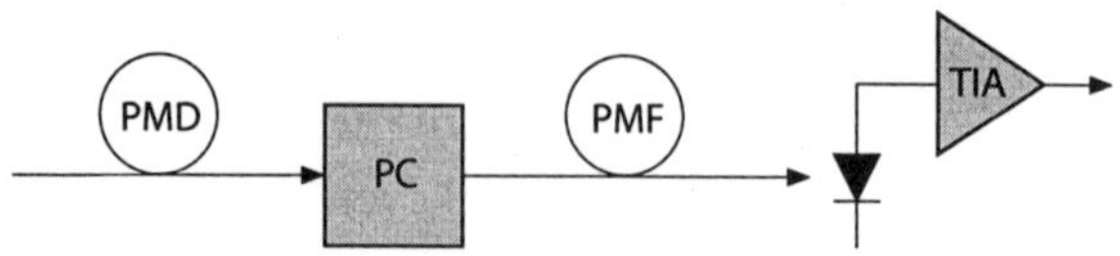

Fig. 4. Simple optical PMD compensator architecture.

any of the optical, optoelectronic and electronic domains. In this section, methods for compensation in each of these domains are described and compared.

3.1. *Optical PMD Compensation*

Optical PMD compensation has been demonstrated to 160 Gb/s[25]. One of the most common optical PMD compensators requires a polarization controller (PC) and a length of polarization-maintaining fiber (PMF), as shown in Figure 4[26]. The PC is used to align the polarization of the light signal such that it is aligned with the the principal states of polarization (PSPs) of the PMF. PMF is fiber which has been intentionally manufactured to have a large, but controlled, birefringence, and therefore can be used to generate a specific amount of DGD. In this way, the power in the fast SOP can be delayed by an amount equal to the DGD of the PMF, resulting in a reduction in the overall DGD. More complicated compensators can be made by replacing the fixed length of PMF with a variable delay to enable cancellation of arbitrary amounts of DGD, or by using multiple PC-PMF stages to increase the degrees of freedom and hence the accuracy of the compensation[12].

Despite the obvious advantages of compensating an optical phenomenon with optical components, optical compensation has several disadvantages. First, optical schemes require expensive and relatively bulky optical components. Also, because of the dynamic nature of PMD, compensators must be adaptive. Adaptation is not easily achieved in the optical domain because of the relative lack of flexibility in optical components, and because of the difficulty in extracting an appropriate error signal to control the adaptation.

3.2. *Optoelectronic PMD Compensation*

It is also possible to compensate PMD using a scheme which involves both the optical and electronic domains. Typically, this scheme involves splitting the received light signal into its two polarization modes by a PC and a polarization beam splitter (PBS)[4]. The resulting light signals are then converted to electrical signals by two separate photodiode-transimpedance amplifier (TIA) front-ends. The electrical signal corresponding to the light in the fast SOP is then delayed by an interval equal to the DGD. Finally, the two signals are recombined to form a received signal that is free from PMD effects. This concept is illustrated in Figure 5.

The main advantage of optoelectronic compensation is that some of the compensation hardware is moved from the optical to the electronic domain, increasing the

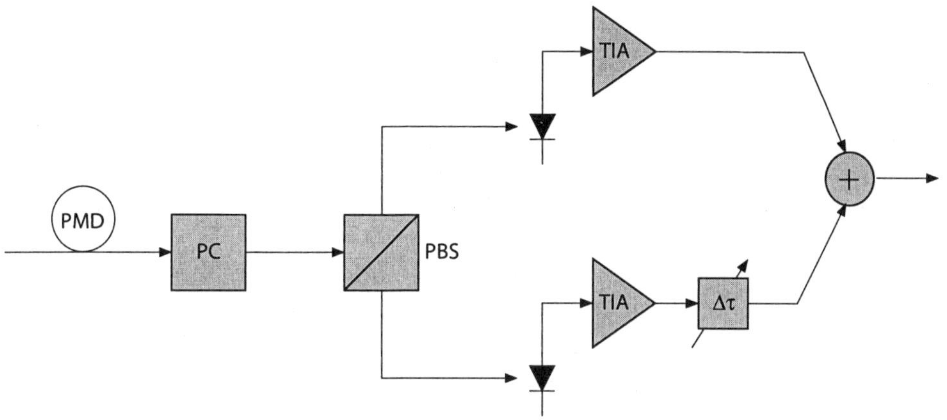

Fig. 5. Typical optoelectronic PMD compensator architecture.

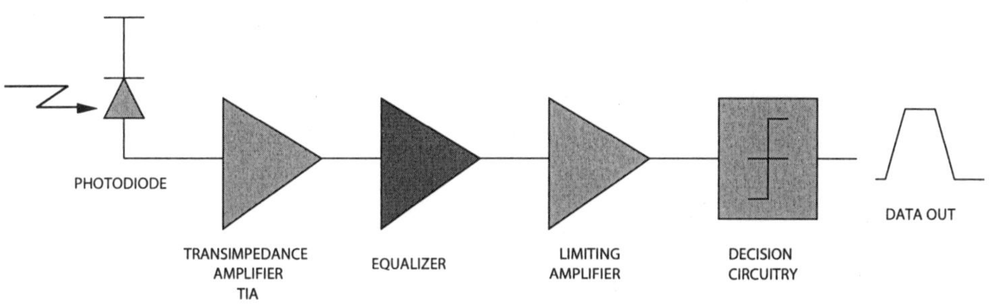

Fig. 6. Block diagram of optical receiver including electronic PMD compensator.

level of integration. However, optoelectronic compensation still requires extra optical components (PC and PBS), so greater integration is possible using an electronic scheme. Also, the addition of a second front-end is a significant expense.

3.3. *Electronic PMD Compensation*

Electronic PMD compensation is performed by equalization of the received signal after it has been converted from light to electricity by a photodiode and TIA. A system diagram of an optical receiver with an equalizer is given in Figure 6.

Electronic equalization is attractive because it offers a higher level of integration and hence a lower cost when compared to optical and optoelectronic solutions. A high level of integration is especially important in WDM systems, in which PMD compensation is required for each channel[4]. Also, the adaptation that is required to track changing PMD conditions is relatively simple to implement electronically, with established adaptation algorithms such as the least mean square (LMS) algorithm readily available. As a result of these considerations, electronic compensation is favoured when it is possible within the bounds of IC technology.

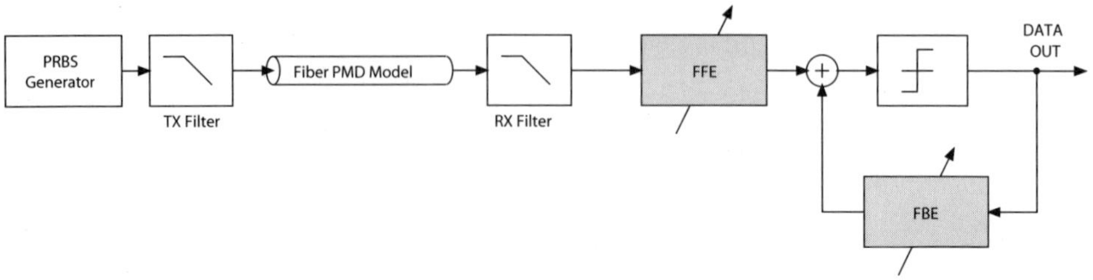

Fig. 7. System model used for Matlab/Simulink simulations.

4. System Model

Matlab/Simulink was used to simulate the effects of PMD in an optical system and to compare various equalizer architectures in terms of their compensation abilities. A simplified block diagram of the system used in these simulations is given in Figure 7. A pseudorandom bit sequence (PRBS) generator is used to generate input data at a rate (R) of 40 Gb/s. This data is passed through a first-order lowpass filter ($f_{3dB} = 0.7 \times R$) used to simulate the effects of the finite bandwidth of the transmitter (TX). It is then passed through an optical fiber model which corrupts the data with PMD. The fiber is modelled using (1). At the output of the fiber model the data is filtered with another first-order lowpass filter ($f_{3dB} = 0.7 \times R$) to simulate the finite bandwidth of the receiver (RX). Equalization is then performed and the equalized waveform is sliced to generate the output data.

Not shown in Figure 7 are the clock recovery and adaptation components of the system. The sampling phase was determined by automatically selecting the clock phase corresponding to the largest eye opening at the output of the channel. Coefficient adaptation for both the FFE and FBE was performed using the LMS algorithm.

Figure 7 shows the equalizer as a DFE, though several equalizer architectures were considered.

5. Equalizer Architectures

5.1. *Analog Equalizer*

Analog or "peaking" equalizers have been used in the past for equalizing simple low-pass channels[27]. The potential advantage of this architecture is its relatively simple implementation. However, the analog equalizer is unsuitable as a PMD compensator because it is not flexible enough to adapt to the wide range of potential PMD conditions. Also, because it is a linear circuit it is unable to compensate for the deep null in the frequency spectrum caused by PMD with γ values near 0.5.

5.2. *IIR Equalizer*

The infinite impulse response (IIR) equalizer would seem to have great potential as a PMD compensator. Because the frequency response of the PMD channel is given by (6), the inverse of the channel transfer function is:

$$H^{-1}_{\mathrm{PMD}}(f) = \frac{1}{\gamma + (1-\gamma)e^{-j2\pi f \Delta\tau}} \tag{8}$$

$H^{-1}_{\mathrm{PMD}}(f)$ also relates the input $X(f)$ and output $Y(f)$ of the inverse filter:

$$H^{-1}_{\mathrm{PMD}}(f) = \frac{Y(f)}{X(f)} \tag{9}$$

Solving for $Y(f)$ we get the input-output relationship:

$$Y(f) = \frac{1}{\gamma}X(f) + \frac{\gamma-1}{\gamma}e^{-j2\pi f \Delta\tau}Y(f) \tag{10}$$

Taking the inverse Fourier transform of (10) yields a difference equation:

$$y(t) = \frac{1}{\gamma}x(t) + \frac{\gamma-1}{\gamma}y(t-\Delta\tau) \tag{11}$$

This difference equation describes an IIR filter. While this architecture would seem to offer perfect (zero-forcing) equalization of a PMD channel, the nature of the feedback loop creates problems in practice. Specifically, for $\gamma \leq 0.5$, the equalizer loop gain, which is equal to $\frac{\gamma-1}{\gamma}$ by inspection of (11) is less than -1, meaning that the equalizer is unstable. Thus, since it is unable to compensate PMD for all values of γ, the IIR filter is unsuitable for implementation as a PMD compensator.

5.3. *FIR Equalizer*

The finite impulse response (FIR) filter is a versatile equalizer architecture which is widely used. FIR filters can, given enough taps, approximate any linear transfer function, making them attractive because of their flexibility. However, the usefulness of an FIR filter as a PMD compensator is severely limited because, as a linear filter, it is unable to compensate for the deep nulls caused by PMD with γ values near 0.5.

5.4. *Decision Feedback Equalizer*

Figure 7 illustrates the basic DFE topology. The DFE consists of an FFE and an FBE, both of which can be implemented as FIR filters for maximum flexibility. The most important advantage of the DFE architecture in terms of PMD compensation is that the use of an FBE introduces nonlinear equalization, allowing compensation of the nulls resulting from γ values near 0.5[11]. Because of this, the DFE is the only architecture surveyed that meets the requirements for an electronic PMD compensator.

The main disadvantage of the DFE is its difficult implementation at high speeds, as a result of the feedback loop inherent to the FBE. However, architectural techniques such as the look-ahead DFE architecture[28,29] can alleviate this problem.

6. Simulation Methodology

Having identified the DFE architecture as the most suitable, simulations were performed to identify the performance tradeoffs with respect to the number of equalizer taps in the FFE and FBE. All simulations were performed with symbol-spaced equalizer taps unless otherwise noted.

For each equalizer configuration, it was necessary to consider a wide range of PMD conditions. The DGD, $\Delta\tau$, was varied from 0 to 100 ps (4 bit periods at 40 Gb/s) and γ was varied from 0 to 1. For each ($\Delta\tau$,γ) pair the equalizer was allowed to converge to the ideal tap weights as determined by the LMS algorithm. Then, the ISI penalty was determined by calculating the amount of eye closure using[30]:

$$\text{ISI penalty (dB)} = 10 \times \log_{10}\left(\frac{max.\ eye\ opening}{min.\ eye\ opening}\right) \tag{12}$$

Figure 8 shows representative eye diagrams for the unequalized and equalized case for one particular ($\Delta\tau$,γ) pair. Figure 9 shows surface plots of the ISI penalty with and without an equalizer for one architecture over a wide range of ($\Delta\tau$,γ) pairs. This plot demonstrates the elimination of the penalty pole at $\Delta\tau = 25$ ps, $\gamma = 0.5$ by equalization with a DFE.

Once the ISI penalty had been calculated for all ($\Delta\tau$,γ) pairs, the cumulative probability (CP) of a system outage given a particular power margin was calculated using[15]:

$$\text{CP} = \sum_{(\Delta\tau,\gamma)'} \rho_1(\Delta\tau)\rho_2(\gamma) \tag{13}$$

where $\rho_1(\Delta\tau)$ is the probability distribution of $\Delta\tau$ as described by (2), $\rho_2(\gamma)$ is the probability distribution of γ (uniform)[15] and $(\Delta\tau,\gamma)'$ is the set of $(\Delta\tau,\gamma)$ pairs for which the ISI penalty is greater than the power margin. Power margin represents the ratio of the transmitted power to the transmitted power required for a given BER (e.g. 10^{-12}). When the ISI penalty exceeds the power margin, a system outage occurs because the excess transmitted power cannot overcome the eye closure caused by the ISI, and the BER increases above the specified maximum tolerable level.

As described in Section 2.1, $\rho_1(\Delta\tau)$ depends on the average DGD of the particular fiber. For each equalizer configuration, the probability distribution was varied by adjusting the average DGD to find the maximum average DGD that would result in a CP of less than 10^{-6} (30 seconds per year).

To summarize, for each equalizer configuration (number of taps) the ISI penalty contour is calculated over all γ and $\Delta\tau$. For a given power margin, the $(\gamma,\Delta\tau)'$ pairs corresponding to a system outage are those pairs that give an ISI penalty greater than the power margin. The joint probabilities of occurrence for all of the $(\gamma,\Delta\tau)'$

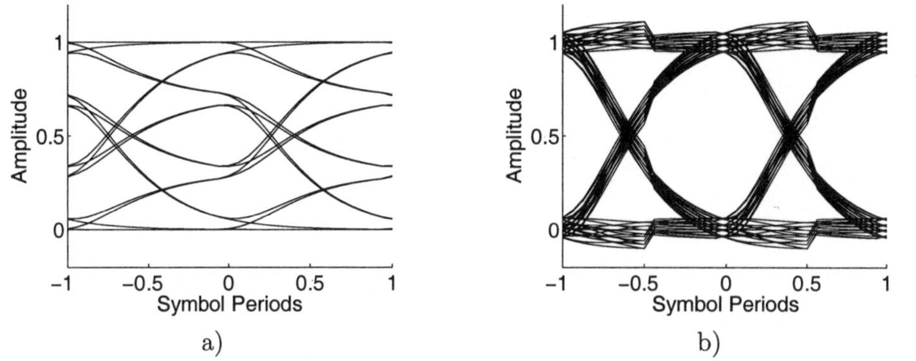

Fig. 8. Eye diagrams for $\Delta\tau = 25$ ps, $\gamma = 0.3$. a) No equalization. b) Equalization by 3-tap FFE and 1-tap FBE.

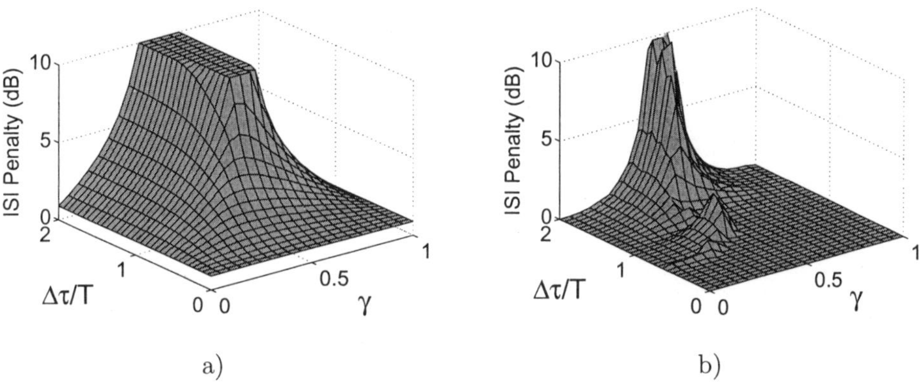

Fig. 9. ISI penalty vs. $\Delta\tau$ and γ. ISI penalty is truncated at 10 dB. a) No equalization. b) Equalization by 3-tap FFE and 1-tap FBE.

pairs contributing to system outage are then summed to find the overall system outage probability. As long as this overall system outage probability remains lower than 10^{-6}, the average DGD is increased and the calculation repeated. The maximum average DGD that results in an overall outage probability of less than 10^{-6} is then recorded as a measure of performance for comparison with other equalizer configurations.

7. Simulation Results

Figures 10, 11 and 12 show the results of these simulations for DFEs with no FBE, a 1-tap FBE and a 2-tap FBE, respectively. In each case, the maximum average DGD that is tolerable from a system point of view is plotted against the power margin for different numbers of FFE taps. In addition, the unequalized case is included as a reference for comparison. Figure 10 shows that using an FFE only, a modest

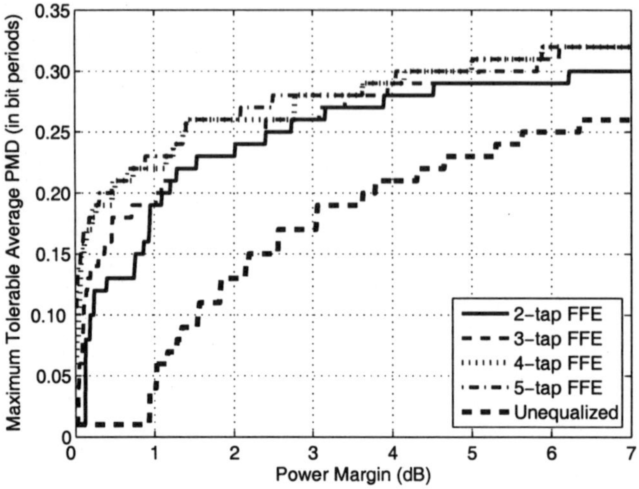

Fig. 10. Plot of maximum tolerable PMD vs. power margin for FFEs with varying number of taps (No FBE).

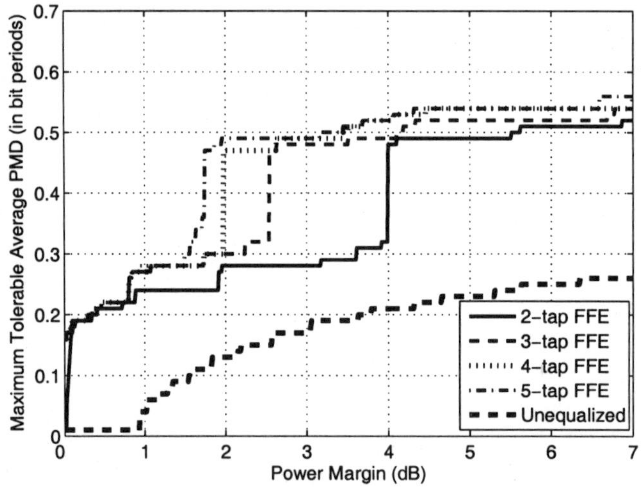

Fig. 11. Plot of maximum tolerable PMD vs. power margin for FFEs with varying number of taps (1-tap FBE).

increase in maximum average DGD is possible, from roughly $0.15T_b$ to $0.25T_b$ at a power margin of 2-3 dB. Only minor improvements are achievable by increasing the number of FFE taps because regardless of the number of taps the FFE is unable to compensate for the case $\Delta\tau = 25$ ps, $\gamma = 0.5$. Figure 11 demonstrates that by using a 1-tap FBE, a significant performance increase is possible, with the maximum

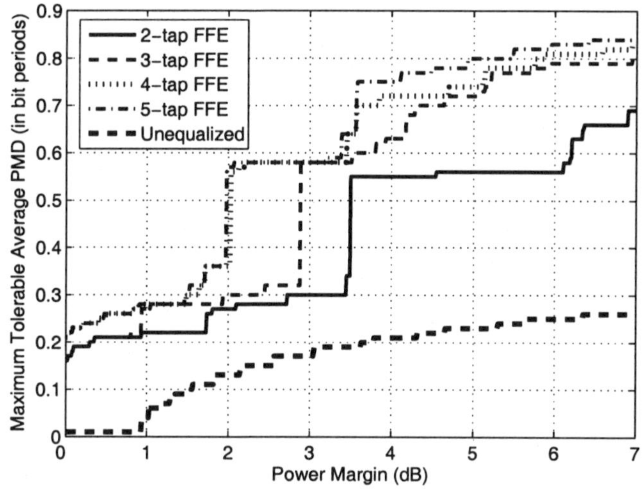

Fig. 12. Plot of maximum tolerable PMD vs. power margin for FFEs with varying number of taps (2-tap FBE).

average DGD increasing to approximately $0.5T_b$. For this case, the number of FFE taps offering the best balance between performance and complexity is dependent on the power margin. For power margins below 3 dB, four taps offer the best balance, while three taps offer the best balance for power margins above 3 dB. Figure 12 shows that a further increase in maximum average DGD is possible by using a 2-tap FBE, but significant gains are limited to power margins above 4 dB. Once again, four FFE taps offer the best balance for power margins below 3 dB, while three taps offer the best balance for power margins above 3 dB.

Figure 13 shows the maximum average DGD plotted against the number of FFE taps for a power margin of 3 dB. Once again, the unequalized case is included for comparison. This plot more clearly shows the performance of each of the equalizer architectures. It is clear from this plot that for a power margin of 3 dB, a 3-tap FFE offers performance nearly equal to the more complex 4- and 5-tap FFEs. As expected, the 2-tap FBE offers a modest performance increase over the 1-tap FBE. However, the 1-tap FBE may be a more attractive choice when this performance increase is weighed against the added complexity of a second tap. This is particularly true if a look-ahead architecture is employed for the FBE (as is likely the case) since the FBE complexity would increase exponentially with the number of taps.

These results are significant because they imply that the useful length of high-speed optical systems affected by PMD can be greatly increased by including electronic PMD compensation in the form of a DFE with modest complexity. To keep system outage levels at an acceptable level, $\Delta\tau_{avg}$ must be less than the maximum average DGD. Therefore, using (4) it is found that the useful length of the fiber increases with the square of the increase in maximum average DGD. As an exam-

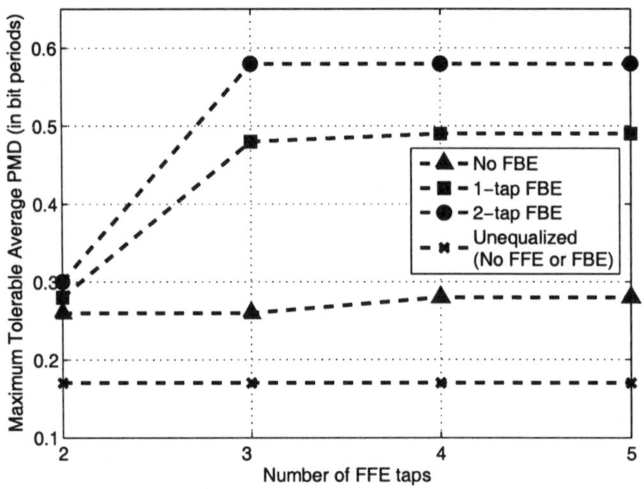

Fig. 13. Plot of maximum tolerable PMD vs. number of FFE taps for no FBE, 1-tap FBE and 2-tap FBE at a power margin of 3 dB.

ple, consider a 40-Gb/s system for which the PMD of the fiber is 1.0 ps/$\sqrt{\text{km}}$, and the power margin is 3 dB. From Figure 13, the maximum average DGD for an unequalized system at a power margin of 3 dB is $0.17T_b$, corresponding to a maximum fiber length of 18 km, using (4). The maximum average DGD for a system using a DFE with a 3-tap FFE and 1-tap FBE is $0.49T_b$, corresponding to a maximum fiber length of 150 km. Hence, an increase in maximum length of more than eight times is achieved by equalization. At these lengths, other impairments (such as noise and CD) would likely replace PMD as reach-limiting factors. Therefore, the combination of a 3-tap FFE and 1-tap FBE is all that is required to effectively eliminate PMD as the dominant length-limiting factor in most 40-Gb/s optical systems.

7.1. *Fractionally-Spaced Equalization*

Fractionally-spaced equalizers (FSEs), for which the tap spacing is $T_B/2$, were also considered for the FFE. Figure 14 shows the maximum average PMD for fractionally-spaced and symbol-spaced FFEs with a 1-tap FBE for a system power margin of 3 dB. The two equalizer configurations are compared in terms of equalizer span, i.e. the difference in delay between the first tap and the last tap of the equalizer. Equalizer span is important in PMD compensation because it determines the maximum amount of DGD that the equalizer can handle. Furtermore, when distributed circuit techniques are employed for implementation of the FFE, equalizer complexity is generally proportional to the span, not the number of taps. From this plot, it is seen that for a given span, the FSE provides a performance increase over the symbol-spaced equalizer. However, the gain through each tap of an FSE

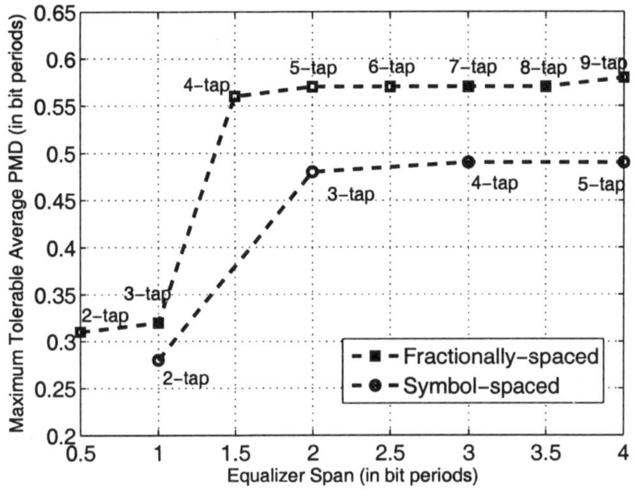

Fig. 14. Plot of maximum tolerable PMD vs. equalizer span for fractionally-spaced and symbol-spaced FFEs and a 1-tap FBE at a power margin of 3 dB.

based upon distributed circuit techniques will be limited to one half that through each tap of a symbol-spaced equalizer because of topological considerations. For cases where noise is the limiting factor and only one equalizer tap is needed (e.g. a long fiber that happens to have negligible PMD), an FSE with only half the tap gain will require a power margin 3 dB higher than a symbol-spaced equalizer for equal performance. Therefore, the performance increase demonstrated in Figure 14 is exaggerated somewhat. Also, increasing the number of taps by decreasing the tap spacing increases the complexity of the adaptation required, making it more difficult to achieve convergence of tap values. For these reasons, a symbol-spaced equalizer is often preferred.

7.2. *Variable Tap Delays*

One observation from these simulations is that the usefulness of a PMD compensator is limited by its span. An equalizer is powerless to compensate for PMD with DGD exceeding its span. To increase the amount of DGD that can be compensated, more taps can be added. However, in a PMD compensator with many taps, only the first taps will be utilized for small DGD values while only the first and last taps will be utilized for large DGD values. Alternatively, an equalizer with only a few taps can perform comparably to one with many taps if the tap delays can be made variable. Variable tap delays allow the tap spacing to be aligned with the DGD, maximizing the compensation accuracy with the fewest possible taps. However, there is currently no acceptable method of implementing variable delays on integrated circuits with the required tuning range and bandwidth.

8. Conclusions

A system-level analysis using Matlab/Simulink has been used to compare the performance of different electronic PMD compensator architectures at 40 Gb/s. It has been demonstrated that equalization by a DFE with a 3-tap FFE and a 1-tap FBE is able to increase by nearly three times the maximum average DGD that is tolerable from a system point of view, from $0.17T_b$ to $0.49T_b$. This is significant because it effectively eliminates PMD as the dominant length-limiting factor in most 40-Gb/s optical systems. In addition to the many practical considerations favoring electronic compensation over optical and optoelectronic methods, the complexity of the required equalizer is modest, with prototype integrated circuits meeting these requirements having already been reported[9].

Acknowledgments

This work has been supported by grants from the Natural Sciences and Engineering Research Council of Canada (NSERC), Gennum Corporation and Micronet.

References

1. J. Hecht, *City of Light: The Story of Fiber Optics.* New York: Oxford University Press, 1999.
2. K. Azadet, E. F. Haratsch, H. Kim, F. Saibi, J. H. Saunders, M. Shaffer, L. Song, and M. L. Yu, "Equalization and FEC techniques for optical transceivers," *IEEE Journal of Solid-State Circuits*, vol. 37, no. 3, pp. 317–327, Mar. 2002.
3. H. Bülow, "Polarisation mode dispersion (PMD) sensitivity of a 10 Gbit/s transmission system," in *Optical Communication, 1996. ECOC '96. 22nd European Conference on*, vol. 2, 1996, pp. 211–214.
4. ——, "PMD mitigation techniques and their effectiveness in installed fiber," in *Optical Fiber Communication Conference, 2000*, vol. 3, 2000, pp. 110–112.
5. H. Bülow, W. Baumert, H. Schmuck, F. Mohr, T. Schulz, F. Kuppers, and W. Weiershausen, "Measurement of the maximum speed of PMD fluctuation in installed field fiber," in *Optical Fiber Communication Conference, 1999, and the International Conference on Integrated Optics and Optical Fiber Communication. OFC/IOOC '99. Technical Digest*, vol. 2, 1999, pp. 83–85.
6. H. Wu, J. A. Tierno, P. Pepeljugoski, J. Schaub, S. Gowda, J. A. Kash, and A. Hajimiri, "Integrated transversal equalizers in high-speed fiber-optic systems," *IEEE Journal of Solid-State Circuits*, vol. 38, no. 12, pp. 2131–2137, Dec. 2003.
7. H. Bülow, F. Buchali, W. Baumert, R. Ballentin, and T. Wehren, "PMD mitigation at 10 Gbit/s using linear and nonlinear integrated electronic equaliser circuits," *Electronics Letters*, vol. 36, no. 2, pp. 163–164, 2000.
8. L. Möller, S. Thiede, S. Chandrasekhar, W. Benz, M. Lang, T. Jakobus, and M. Schlechtweg, "ISI mitigation using decision feedback loop demonstrated with PMD distorted 10Gbit/s signals," *Electronic Letters*, vol. 35, no. 24, pp. 2092–2093, 1999.
9. M. Nakamura, H. Nosaka, M. Ida, K. Kurishima, and M. Tokumitsu, "Electrical PMD equalizer ICs for a 40-Gbit/s transmission," in *Optical Fiber Communication Conference and Exhibit, 2004. OFC 2004*, vol. TuG4, 2004.
10. A. Hazneci and S. P. Voinigescu, "A 49-Gb/s, 7-tap transversal filter in 0.18μm SiGe

BiCMOS for backplane equalization," in *IEEE Compound Semiconductor Integrated Circuits Symposium*, Monterey, CA, Oct. 2004.
11. H. Bülow and G. Thielecke, "Electronic PMD mitigation - from linear equalization to maximum-likelihood detection," in *Optical Fiber Communication Conference and Exhibit, 2001. OFC 2001*, vol. 3, 2001, pp. WAA3–1–WAA3–3.
12. H. Sunnerud, C. J. Xie, M. Karlsson, R. Samuelsson, and P. A. Andrekson, "A comparison between different PMD compensation techniques," *IEEE/OSA Journal of Lightwave Technology*, vol. 20, no. 3, pp. 368–378, Mar. 2002.
13. B. Razavi, *Design of Integrated Circuits for Optical Communications.* New York: McGraw-Hill, 2003.
14. C. D. Poole, R. W. Tkach, A. R. Chraplyvy, and D. A. Fishman, "Fading in lightwave systems due to polarization-mode dispersion," *IEEE Photonics Technology Letters*, vol. 3, no. 1, pp. 68–70, Jan. 1991.
15. H. Bülow, "Operation of digital optical transmission system with minimal degradation due to polarisation mode dispersion," *Electronic Letters*, vol. 31, no. 3, pp. 214–215, 1995.
16. G. J. Foschini and C. D. Poole, "Statistical-theory of polarization dispersion in single-mode fibers," *IEEE/OSA Journal of Lightwave Technology*, vol. 9, no. 11, pp. 1439–1456, Nov. 1991.
17. N. Gisin, R. Passy, J. C. Bishoff, and B. Perny, "Experimental investigations of the statistical properties of polarization mode dispersion in single mode fibers," *IEEE Photonics Technology Letters*, vol. 5, no. 7, pp. 819–821, July 1993.
18. R. Ramaswami and K. N. Sivarajan, *Optical Networks - A Practical Perspective, Second Edition.* Morgan Kaufmann Publishers, 2002.
19. F. Buchali and H. Bülow, "Adaptive PMD compensation by electrical and optical techniques," *IEEE/OSA Journal of Lightwave Technology*, vol. 22, no. 4, pp. 1116–1126, Apr. 2004.
20. B. Wedding and C. N. Haslach, "Enhanced PMD mitigation by polarization scrambling and forward error correction," in *Optical Fiber Communication Conference and Exhibit, 2001. OFC 2001*, vol. 3, 2001, pp. WAA1–1–WAA1–3.
21. L. Song, M.-L. Yu, and M. S. Shaffer, "10- and 40-Gb/s forward error correction devices for optical communications," *IEEE Journal of Solid-State Circuits*, vol. 37, no. 11, pp. 1565–1573, Nov. 2002.
22. D. Penninckx, S. Lanne, and H. Bülow, "WDM redundancy to counteract PMD effects in optical systems," in *Optical Communication, 2001. ECOC '01. 27th European Conference on*, vol. 3, 2001, pp. 444–445.
23. H. F. Haunstein, W. Sauer-Greff, A. Dittrich, K. Sticht, and R. Urbansky, "Principles for electronic equalization of polarization-mode dispersion," *IEEE/OSA Journal of Lightwave Technology*, vol. 22, no. 4, pp. 1169–1182, Apr. 2004.
24. H. Sunnerud, M. Karlsson, C. J. Xie, and P. A. Andrekson, "Polarization-mode dispersion in high-speed fiber-optic transmission systems," *IEEE/OSA Journal of Lightwave Technology*, vol. 20, no. 12, pp. 2204–2219, Dec. 2002.
25. H. Sunnerud, M. Westlund, J. Li, J. Hansryd, M. Karlsson, P.-O. Hedekvist, and P. A. Andrekson, "Long-term 160 Gb/s-TDM, RZ transmission with automatic PMD compensation and system monitoring using an optical sampling system," in *Optical Communication, 2001. ECOC '01. 27th European Conference on*, vol. 6, 2001, pp. 18–19.
26. C. Francia, J.-P. Bruyère, F. Thiéry, and D. Penninckx, "Simple dynamic polarisation mode dispersion compensator," *Electronics Letters*, vol. 35, no. 5, pp. 414–415, 1999.
27. G. Zhang, P. Chaudhari, and M. Green, "26.7 - a BiCMOS 10Gb/s adaptive ca-

ble equalizer," in *Solid-State Circuits Conference, 2004. Digest of Technical Papers. ISSCC. 2004 IEEE International*, 2004, pp. 482–491.

28. S. Kasturia and J. H. Winters, "Techniques for high-speed implementation of non-linear cancellation," *IEEE J. Selected Areas in Communications*, vol. 9, no. 5, pp. 711–717, June 1991.
29. Y.-S. Sohn, "A 2.2 gbps CMOS look-ahead dfe receiver for multidrop channel with pin-to-pin time skew compensation," in *IEEE Custom Integrated Circuits Conference*, Sept. 2003, pp. 473–476.
30. P. Pepeljugoski, S. E. Golowich, A. J. Ritger, P. Kolesar, and A. Risteski, "Modeling and simulation of next-generation multimode fiber links," *IEEE/OSA Journal of Lightwave Technology*, vol. 21, no. 5, pp. 1242–1255, May 2003.

International Journal of High Speed Electronics and Systems
Vol. 15, No. 3 (2005) 567–579

Trade-offs in High-Speed Serial Link ICs

Shenggao Li*
Intel Corporation
Fremont, CA 94538/USA
† *lisg@ieee.org*

Serial IOs are widely used to expand the system bandwidth in communication systems. This paper provides an overview of serial IO design trade-offs with regard to power, cost, and performance. Circuit techniques are discussed to achieve low jitter and high bandwidth.

Keywords: SerDes, Serial IO, Clock Generation, Clock and Data Recovery (CDR), High-speed Transceiver, Multi-phase, Jitter, Broadband.

1. Introduction

A wide range of applications nowadays use serial interfaces to achieve high-speed data transmission. Table 1 is a brief summary on some of the popular applications involving serial data communications, all of which feature an embedded clocking scheme, and differential signaling (electrical interface) to achieve high-speed transmission. In comparison with early parallel data bus technologies such as PCI, PCIX, or parallel ATA, a serial link is less susceptible to crosstalk, ground bounce, and clock skew because of the embedded clocking scheme and differential signaling, therefore can achieve much higher data rates over a considerably longer transmission distance. To this date, high-speed serial I/Os have emerged to become a part of routine design in CMOS ICs for high speed high bandwidth computing and communication systems such as computer chipsets and network processors.

Key design requirements for high-speed IOs are low power and high-performance (for example, jitter, sensitivity and BER) with low system cost. Among the different serial applications in Table 1, SONET/OC192 has the most stringent jitter specifications. Fig.1 shows the relative evolution of a 10Gbps SONET SerDes with regard to die area, power and technology. From separated implementation (a) to full integration (b), there is no obvious saving in die area, due to the fact that the die area is pad limited. The saving in power and die area is only obvious when

*44235 Nobel Dr., Fremont, CA 94538

Table 1. Some representative high-speed serial interfaces

Standard	**Data Rate (Gbps)**	**Description**
SONET	9.95	Telecomm, OC192
OC192 over FEC	10.71	ITU G.709
10GE	10.31	802.3ae Ethernet, Datacomm for WAN/MAN
10GE over FEC	11.09	ITU G.709
10G Fiber channel	10.52	Storage area network
XFI	multi-rate	Hot pluggable chip-level electrical interface with small footprint supporting OC192/10GE and 10G fiber channel; FR4 8 12 inches
SATA (G1/G2/G3)	1.5/3.0/6.0	Internal storage devices attachment. Copper (up to 1m)
XAUI	3.125	A low cost medial extender in 10GE application. 4 lanes to provide an aggregate rate of 10Gbps. FR4 (24 inches)
PCI-Express (G1/G2)	2.5/5.0	General purpose I/O, allow multiple lanes.
InfiniBand	2.5/5/10	Data-center (server) interconnect, allow multiple lanes, support copper (17m) and optical cable (10km).

higher level of integration is achieved where inter-chip data buses are eliminated in (c) and new technology is adopted.

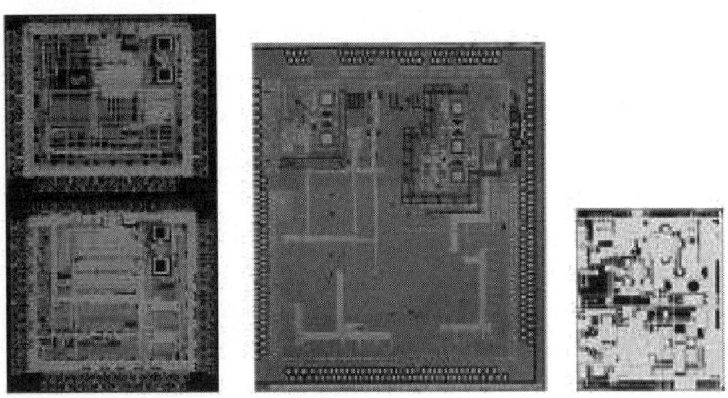

(a) (b) (c)

Fig. 1. 10Gbps SerDes – design evolution. (a) Separate TX/RX (0.5W each, .13um) ; (b) Fully integrated SerDes (1W, .13um); (c) SerDes Macro (< 0.5W, 90nm)

The saving could be at the expense of circuit performance. As an example, when high-speed IOs are integrated with a noisy digital processing core in a low cost CMOS process, the jitter performance of high-speed IOs will suffer from high switching noise through substrate coupling. In another aspect, in 10 Gbps backplanes or other applications, the channel (copper cable or PCB trace) over which data is transmitted is very lossy and band-limited. When signal reaches the receiver, the signal eye will be highly distorted. These are design challenges that need to be tackled at circuit level, architecture level, and system level.

This paper will discuss the design trade-offs of high-speed serial link interfaces. In the following discussion, section 2 provides an overview on system partitioning from a power and cost perspective. Section 3 discusses clocking strategy at architecture level, followed by discussions of clock generation schemes in section 4. Section 5 is dedicated to broadband channel design. A CML driver with programmable bandwidth is given. Conclusions are made in section 6.

2. System Partitioning

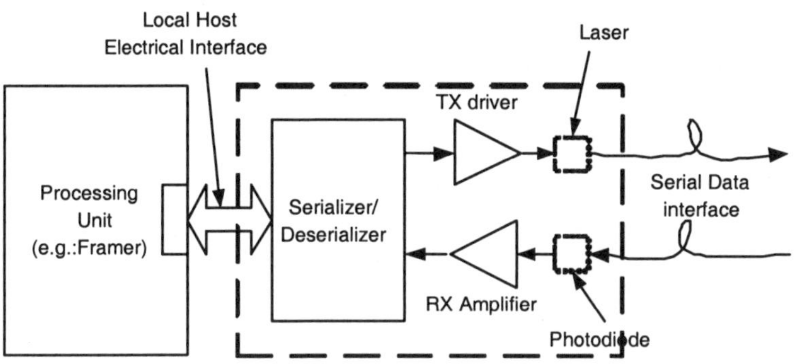

Fig. 2. transceiver block diagram in a serial interface

Fig.2 is the transceiver block diagram of a typical serial interface. This architecture applies to all of the serial interfaces listed in Table 1, with the exception that additional electrical/optical conversion devices are required in optical links, as identified by the laser (E/O) and the photodiode (O/E). Optical serial links offer the highest link performance, but are still expensive due to high cost in those optical components, for two reasons: Firstly, the laser and the photo detector are typically made with exotic material such as GaAs, AlGaAs, or InP, which are expensive and difficult for high-yield production and large-scale integration. Secondly, for the laser and photo detector to operate stably, proper optical packaging technologies are required to handle issues such as laser cooling, optical alignment, and signal integrity. Manufacturing cost, yield, and reliability remain critical in driving down the overall cost of optical modules.

The core function of the serial IO in Fig.2 is represented by the SerDes, of which the Serializer is composed of a clock generator (PLL) and an N:1 data multiplexer, and the Deserializer is composed of a clock and data recovery circuit (CDR) and a 1:N data de-multiplexer. Various features such as error detection, loop-back test, and PRBS pattern generation can be integrated with the SerDes core for diagnosis purpose in a fully integrated transceiver. CMOS technologies provide the cost and power benefit while providing those complex functionalities.

The transmit driver and the receive amplifier are more application dependent. Very often they may not be feasible for CMOS integration. For instance, for serial interfaces such as XFI, PCIE and SATA where the driver power and receiver sensitivity requirement are relatively relaxed, it is sufficient to have the TX driver and the RX amplifier integrated with the SerDes in CMOS processes. On the other hand, for optical communications especially long reach telecomm or metropolitan area network where the TX driver and the RX amplifier need to achieve high current swing and high input sensitivity respectively, the direct laser driver and part of the input amplifier (TIA) are designed with InGaAs, GaAs or SiGe.

To optimize system cost, power and performance, a few areas can be looked at:

(1) In the optical communication domain, there is still a lot of room for new technologies to bring down system cost, such as tunable laser, and low power and small form factor optical modules, as exemplified by the development of various form-factor optical packages (SFF, SFP, XENPAK, XPAK/X2, and XFP).
(2) Instead of turning to high-cost fiber link for high-quality data transmission within short distance such as the backplane, currently there are considerable efforts on 10Gbps backplane study, including channel characterization, signaling and transceiver architecture.
(3) Third, the cost saver on the IC side is to integrate the high-speed SerDes into the processing unit (processors and switch box) such that the parallel electrical interface is eliminated (Fig.2). To this date, there are already significant work and products that embed multiple high-speed (up to 10Gbps) IOs such as PCIE, SATA, and XFI.

Design challenges (such as power distribution, power dissipation, package routing, and noise isolation) may arise when high-speed IOs are integrated with the digital processing core. Without careful evaluation, the cost will be driven up with degraded system performance.

One way of getting around the problem is illustrated in Figure 3, which is also the primary design concept of the XFI interface(refer to Table 1).[2] In this scheme, one or multiple high-speed SerDes's (HSIO) of up to 10 Gbps are integrated with the processing unit. For each SerDes, one additional off-chip signal conditioner is used to reshape and regenerate signal to and from the HSIO. The benefits are as follows:

(1) With on-chip integration, the parallel interface (C in Fig.3) of the HSIO is now becoming on-die interconnections. This saves significantly pin counts and power consumption which are otherwise required by inter-chip parallel data buses.
(2) The 1:1 CDR in the signal conditioner relaxes the design requirement on the

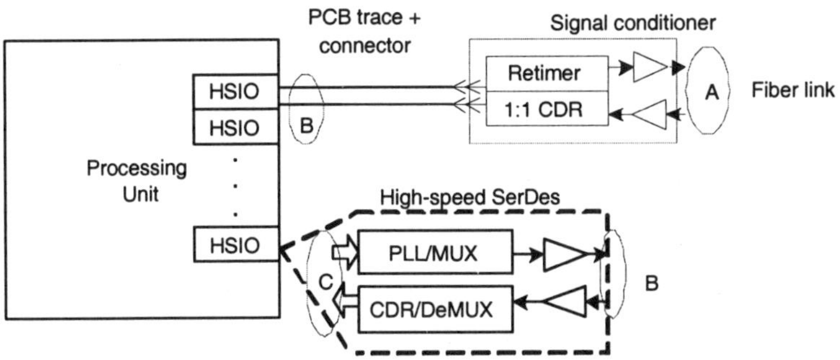

Fig. 3. Integrated SerDes with off-chip signal conditioner

HSIO receiver, because the weak data input from the fiber link is now amplified and regenerated by the 1:1 CDR. Although signal degradation from the PCB trace and connector can be significant due to electrical loss and cross talk, the quality of the regenerated signal is still much better at point B than the received signal at point A. For example, the required input sensitivity at point B is relaxed to 110 mV differential in XFI, while the input sensitivity at point A for SONET is typically 10 mV differential. This greatly reduces the receiver design requirement of the HSIO receiver in terms of noise isolation between channels, amplifier gain and overall power consumption.

(3) Similarly, the re-timer block in the signal conditioner relaxes the design requirement on the HSIO transmitter. The re-timer typically use a clock recovery circuit or a high-quality clock generator to clean up the jitter by the HSIO transmitter and the electrical channel. For example, the total jitter specified at point B in XFI is 0.30 UIpp ($\sim$ 30 ps, per unit interval is 100 ps for a 10 Gbps data rate) with the re-timer, the jitter at point A is reduced to 0.1 UIpp ($\sim$ 10 ps) to satisfy SONET specification. The relaxed specification at point B reduces the design requirement of the HSIO transmitter in terms of clock quality, driver bandwidth, and overall power consumption.

(4) It is true that the signal conditioner consumes additional power, however, the serial XFI interface has reduced the complexity and form factor of the optical modules with significant saving in power (max power for XENPAK is 11W, XPAK/X2 is 5W, and XFI is 3W. [1]).

(5) As a result of (2) and (3) above, an HSIO could be implemented with low quality components without using expensive mixed-signal process features in a low cost CMOS technology. This cost saving can be significant.

3. SerDes - Architecture Consideration

While fully integrated CMOS SerDes's (transceiver) at 10 Gbps rate have been production proven in the past 2-3 years, there are still considerable challenges for designers. In particular when it comes to full integration (stand-alone transceiver, or integrated transceiver as a high-speed IO of a more complicated IC system), power and area reduction, and noise isolation are the interested topics.

In a traditional transceiver architecture, as shown in Fig.4, the transmitter PLL and the receiver CDR are two independent loops, with the transmit VCO and the receive VCO sitting on the same die. Extra caution is needed in layout to prevent the two VCOs from interacting with each other through substrate and power distribution network. So, typically the two VCOs are located farther apart in a die to ensure jitter performance. In this transceiver architecture, a frequency detector (FD) is used in the RX CDR to continuously monitor the frequency difference between the VCO and the embedded clock in the input data stream. If the difference is more than a predetermined threshold, an auxiliary phase-frequency detector (PFD) is selected such that the CDR loop operates as a regular PLL to acquire frequency locking. Once the frequency difference is within the threshold, the phase detector (PD) is switched in and the CDR is in normal data acquisition. The additional FD and PFD incur additional current consumption, although the current consumption may be minimized by operating the FD and PFD at lower frequency.

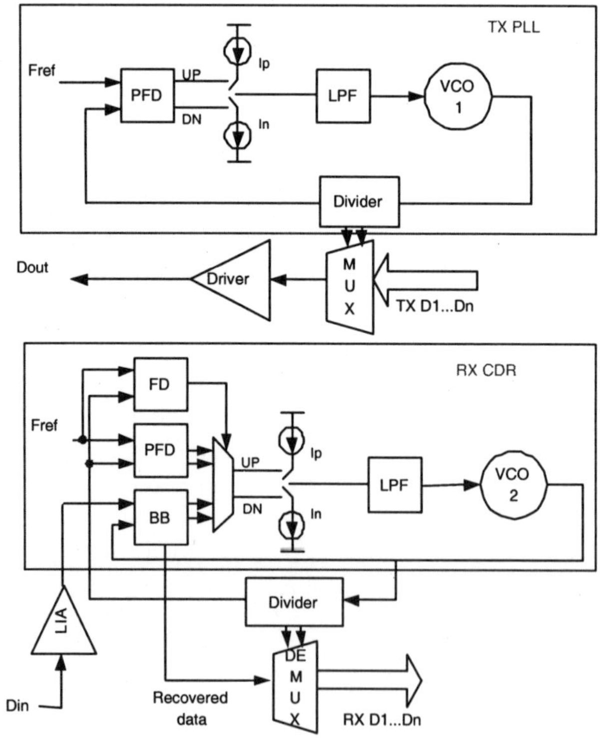

Fig. 4. dual-VCO transceiver architecture (conventional)

In order to get around of the cross talk issue, one may use a phase interpolation scheme for the CDR, such that the CDR clock information is derived from the TX VCO, and no RX VCO is needed.[34] To achieve this goal, a TX PLL with multi-phase clock generation is needed. A phase interpolator is then used to derive the correct sampling clock edges by comparing and adjusting the interpolated clock edges with the transitioning edges of the input data stream. A phase-interpolation based transceiver architecture is shown in Fig.5. Because one VCO is shared between the transmitter and the receiver, there is no frequency pulling issue between two VCOs of the conventional architecture. For most communications systems, the TX clock frequency at the data source and RX clock frequency (derived from the local TX) at the data sink are within a few hundred *ppm*, so as long as the loop bandwidth of the CDR is wide enough, the interpolated clock rate will be able to catch up with the incoming clock rate. There is no need for auxiliary locking mechanism such as the FD and PFD used in Fig.4.

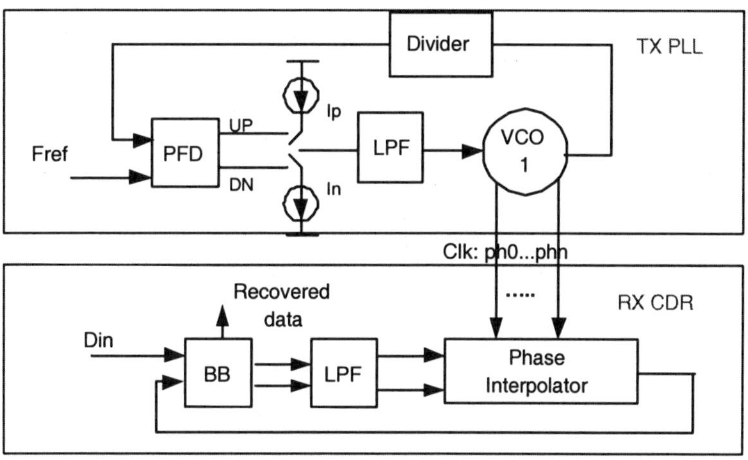

Fig. 5. CDR based on phase-interpolation

The disadvantage is that at frequencies as high as 10 GHz, the phase interpolator can be very power consuming. Achieving good phase linearity is another problem. To alleviate these problems, K phase interpolators may be used to generate $K * 2$ equally spaced sampling clocks simultaneously. The operating frequency of each phase interpolator is then reduced to F_{rate}/K accordingly, where F_{rate} is the input data rate.[3]

4. Clock Generation

This section moves to circuit-level design with focus on multi-phase clock generation using quadrature LC-tank based VCO.

A multiple-stage ring oscillator can be used in Figure 5 to generate multi-phase clocks. However, in jitter sensitive applications such as 10 Gbps SONET, the jitter performance of a ring oscillator does not meet the requirement. Alternatively, a delay locked loop (DLL) can be used along with a regular VCO in the transmitter to generate multi-phase clocks. In comparison with a multi-stage ring oscillator, a DLL offers better jitter performance because jitter doesn't accumulate in the delay cells. On the other hand, a DLL will add a third loop (and hence large loop filter cap) to the system which already has two loops: the TX PLL loop and the RX CDR loop. Specific care is needed to avoid harmonic locking in a DLL [5]. For high clock frequency, the delay line needs to have a high bandwidth in order to maintain constant swing across the delay line, and thus consumes more power.

A quadrature LC-VCO (QVCO) is an idea candidate to provide outstanding jitter performance while consuming low power.[6] Fig.6 shows the topology of a QVCO, in which two regular LC-VCO cores are mutually coupled to obtain four clock phases having phase spacing of 90 degree.

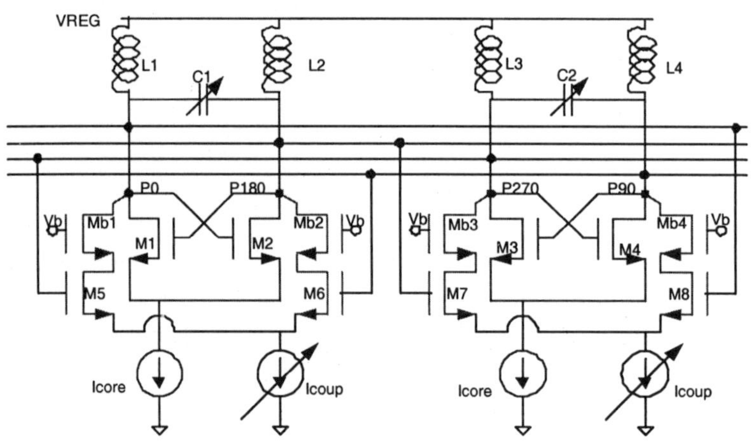

Fig. 6. A quadrature LC-VCO

The QVCO is quite similar to a two-stage ring oscillator (Fig.7), in which the typical delay cell is now replaced by a self-oscillating LC-VCO. The ring structure ensures that X and Y are 90 degree phase apart. It is noted that in a regular ring oscillator, the delay cell may be treated as a low-pass filter, while in a QVCO the delay cell may be treated as a band-pass filter. This treatment leads to a better understanding of the added complexities in designing a QVCO:[6] One issue is that mathematically the QVCO may operate stably at two different frequencies - referred as bi-modal oscillation. The transistors Mb1-4 in Fig.6 were used to resolve bi-modal oscillation, as described in.[6] A second issue is that the coupling between the two VCO cores (Icoup) needs to be large enough to main single tone oscillation. A third

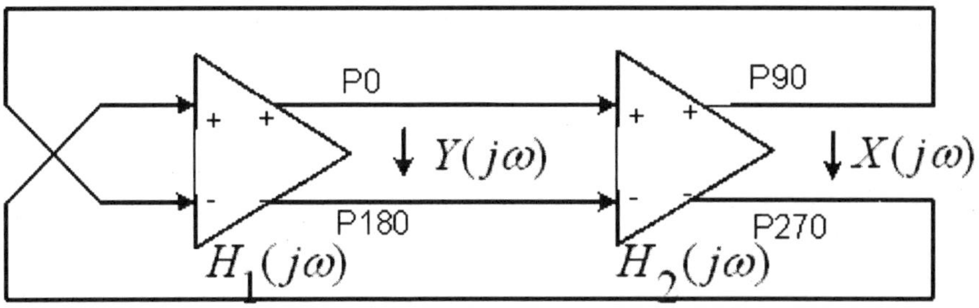

Fig. 7. Even-stage ring-oscillator

issue is the QVCO has a much smaller tuning range than a regular ring-oscillator, and hence requires the use of digitally tuned capacitor array (C1 and C2 in Fig.6) to extend the tuning range. Further, to easy the effort in product testing and trimming, an automated frequency tuning method has been implemented to control the capacitor array. A block diagram of the automated tuning is shown in Fig.8, in which the frequency steering logic is implemented by a frequency comparator and a successive approximation algorithm is used which provides fast frequency locking.

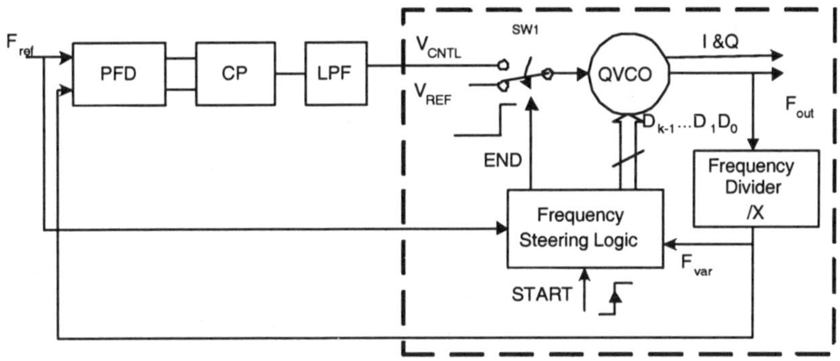

Fig. 8. An automated frequency tuning method

5. Broad-Band Channel

Previous discussions focus on clocking schemes that provide a low-jitter clock base for high-quality data transmission and data recovery. There are significant limitations on the signal channel that also impact the performance of high-speed IOs. Fig.9 illustrates the primary components of a high-speed channel. It is noted that the TX_data which is re-timed by the transmit clock (TX_CLK) is free of jitter to the degree of the achievable TX_CLK quality. The signal goes through all kinds of degradation (distortion, attenuation, interference, and refection, and so on) when

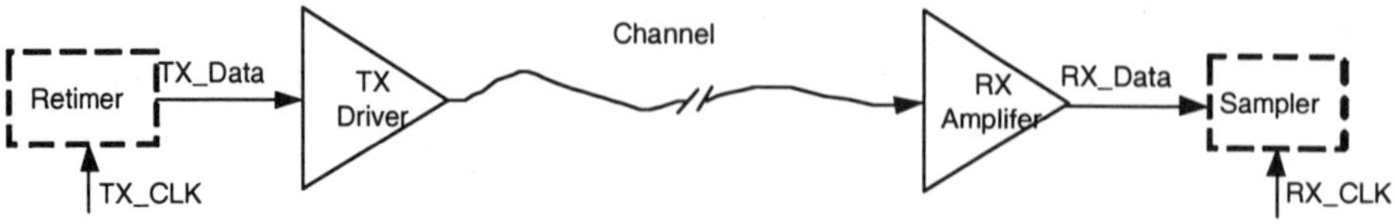

Fig. 9. Signal Channel

passing through the TX driver, the physical channel, and the RX amplifier. By the time it reaches the sampler on the receiver side, the received data eye may become quite closed or even collapsed, resulting in error sampling by the receive clock (RX_CLK) and thus high bit error rate (BER).

A close look at each of the detrimental channel effects are discussed below to evaluate the measures for reliable data transmission:

(1) Circuit bandwidth: most of today's high-speed serial interfaces use NRZ (non-return-to-zero) signaling for data transmission because of the less complexity in transceiver circuit implementation. However, for a random NRZ data stream, the power spectrum will follow the sinc function, as given by $G(f) = E_bT_p(\frac{sin(\pi f T_p)}{\pi f T_p})^2$, where T_p is the bit period, and E_b is the average energy per transmitted bit. This shows that the energy of the data stream occupies a broad frequency range. The limited bandwidth of a communication channel will remove the high-frequency content of the signal, thus causing inter-symbol interference (ISI) jitter. It is calculated that 94% of the total energy is contained between $(0 \sim 0.75)/\mathrm{Tp}$ in a random NRZ data.[7] Therefore, to avoid band-limiting the signal, the bandwidth of the TX driver and the RX amplifier needs to be above $0.7 \sim 0.8$ GHz in a 10 Gbps application. Broadband design techniques based on feedback theory,[8] passive or active inductive peaking may be used to achieve the required bandwidth, along with the scaling on CMOS technologies. Two issues may arise in the pursuit for bandwidth: firstly, a too wide bandwidth will hurt the signal to noise ratio (SNR) because excessive noise passes through to the receiver; secondly, excessive peaking may extend the bandwidth, but hurt the group delay and therefore ISI jitter.
(2) Channel loss: typical communications channels are based on twisted pair, PCB trace, coaxial cable, or optical fiber. Electrical magnetic emission, resistive loss, dielectric absorption, and skin effect are the primary impairments in an electrical channel. Optical fiber doesn't suffer from these impairments and thus allow long distance transmission, however, light signal will attenuate and distort through optical absorption and Rayleigh scattering and wave dispersion. Fig.10 shows the impairment of a 10 Gbps signal being transmitted over an FR4 channel. As the channel length is doubled (16-in vs 8-in), the signal eye diagram is completely closed. The eye closure is caused by the excessive attenuation of high-frequency content in the signal. Simple circuit techniques employing pre-

emphasis on the transmitter side, and post equalization in the receiver side are effective in re-opening the eye by suppressing low frequency content while boosting high frequency content. In more complicated cases, where the channel experience unpredictable spectrum loss and discontinuity due to connector or via impairment, adaptive equalization with multiple-tap filters may be used.

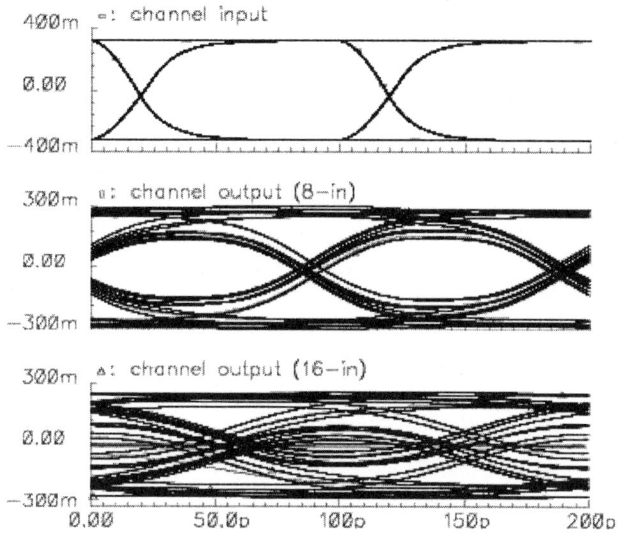

Fig. 10. Channel loss in an FR4 channel

(3) Cross talk: For system with high input sensitivity and large output signal swing, good isolation is needed to ensure that the transmitted signal doesn't corrupt the input signal eye. Channel to channel interference is another concern when multiple channels are grouped together and interfere with one another over cheap media (e.g.: twisted pair) and connectors. Better design techniques on the connectors and better isolation and shielding on the channel are the keys for cross-talk reduction. Additionally, cross-talk cancellation might be considered if improvement on the physical media is restricted.

(4) Termination and reflection: Termination mismatch or channel discontinuity is another source of signal degradation. Many of the high-speed IOs have integrated on-chip 50 Ohm termination resistors with the ability for automatic trimming. At high-frequencies, however, the benefit of the trimming scheme may diminish since it introduces high capacitance at the termination nodes, consequently reducing the effective impedance, as well as the circuit bandwidth.

(5) Electrical magnetic interference (EMI): While it is desirable to have a data eye as open as possible for low BER in the receiver, the fast data transition will cause high EM emission harmful to the environment. Fast transition is also

a primary source for cross talk because of the high-frequency content in the signal. Simply put, EMI needs to be reduced to the minimum level by reducing signal rise/fall time, reducing signal amplitude, or providing good shielding.

Finally, Fig.11 illustrates a broadband driver with bandwidth control where the resistive feedback is used to extend the bandwidth,[8] whereas the capacitive feedback is used to reduce the bandwidth for EMI compliance in EMI critical systems.

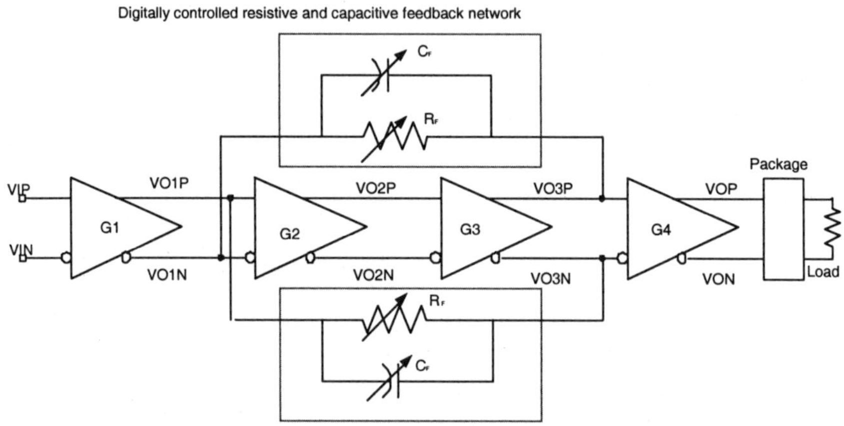

Fig. 11. A broadband driver

6. Conclusion

This paper examined the design trade-offs of high-speed serial IOs over a wide scope ranging from system partitioning, architecture selection, to circuit design techniques, with emphasis on jitter reduction, power and cost saving.

As semiconductor technologies continue to shrink in feature size, the number of transistors per integrated circuit and hence the processing power of integrated circuits has continued its trend according to Moore's Law.[9] The bandwidth limitation of peripheral IOs is now becoming outstanding. High-speed serial IOs provide the flexibility and scalability to extend the IO bandwidth, and further, to extend the processing power of traditional processors through parallel processing.

New serial IO technologies catering to various system requirements continue to shrive, with the ultimate goal of cramming more information onto limited channels of limited bandwidth using minimum power.

Acknowledgment

The author is grateful for many of the discussions with Fulvio Spagna, Peter Kwok, Rui Song, Lidong Chen at Intel Corporation.

References

1. Pierre Herve, Shlomo Ovadia, "Optical Technologies for Enterprise Networks", *Intel Technology Journal*, May, 2004.
2. "10Gigabit Small Form-factor Pluggable Module", *XFP rev3.1*, www.xfpmsa.org.
3. M. Lee, W. Dally, et al, "An 84-mW 4-Gb/s Clock and Data Receovery Circuit for Serial Link Applications", in *Symp. VLSI Circuits*, June 2001, pp. 149-152.
4. Patrik Larsson, "A 2-1600MHz 1.2-2.5 V CMOS clock-recovery PLL with feedback phase-selection and averaging phase-interpolation for jitter reduction", in *ISSCC* 1999, pp356-357.
5. Ching-che Chung, Chen-Yi Lee, "A New DLL-Based Approach for All Digital Multiphase Clock Generation", in *J. of Solid-state Circuits*, vol39, March 2004, pp. 469-475.
6. Shenggao Li, Issy Kipnis, Mohammed Ismail, "A 10GHz CMOS Quadrature VCO for Multi-rate Optical Applications", in *IEEE J. of Solid-state Circuits*, vol38, Oct 2003, pp1626-1634.
7. "NRZ bandwidth, HF cut off vs. SNR", *Maxim Application Notes HFAN-09.0.1.*
8. C. D. Holdenried, J. Haslett, M. Lynch, "Analysis and Design of HBT Cherry-Hooper Amplifiers with Emitter-Follower Feedback for Optical Communications", in *IEEE J. of Solid-state circuits*, Vol39, Nov. 2004, pp.1959-1967.
9. Gorden Moore, "Cramming more Components onto Integrated Circuits", in *Electronics*, vol8, April, 1965.

International Journal of High Speed Electronics and Systems
Vol. 15, No. 3 (2005) 581–597

World Scientific
www.worldscientific.com

HIGH-SPEED ARCHITECTURES AND BUILDING BLOCKS FOR CLOCK AND DATA RECOVERY SYSTEMS

CICERO S. VAUCHER
Philips Research Labs, Prof. Holstlaan 4,
5656AA Eindhoven, Netherlands
cicero.vaucher@philips.com

JAN-PETER FRAMBACH
Philips Semiconductors,
Nijmegen, Netherlands
jan-peter.frambach@philips.com

JOHAN D. van der TANG
Eindhoven University of technology,
Eindhoven, Netherlands
j.d.v.d.tang@tue.nl

This paper focuses on high performance architectures and building blocks for clock and data recovery (CDR) applications. After a review of basic concepts, a parallel CDR architecture for high operation speeds and low-power dissipation is introduced, followed by discussion of linear detectors for multiple octave operation CDR systems. A frequency acquisition architecture, based on standard PLL building blocks, is then presented. Finally, multi-phase and wide-tuning-range oscillators are discussed.

Keywords: Clock-data recovery; phase-locked loop; phase detector; frequency detector; voltage-controlled oscillator; multi-phase oscillator.

1. Introduction

The purpose of a clock and data recovery (CDR) circuit is to extract a clock from an incoming data stream and sample the received data with the extracted clock, so as to be further processed in a synchronous manner. A CDR thus performs two tasks, clock recovery from a raw data stream and (bit) synchronization of the raw data stream. The retimed data stream is now suitable for further signal processing, such as byte alignment, frame synchronization, overhead processing, payload extraction and overhead based system performance monitoring.[1] Although several ways exist to perform clock recovery, for example based on non-linear processing and SAW filters, monolithic implementations are usually based on a Phase Locked Loop (PLL).[2,3]

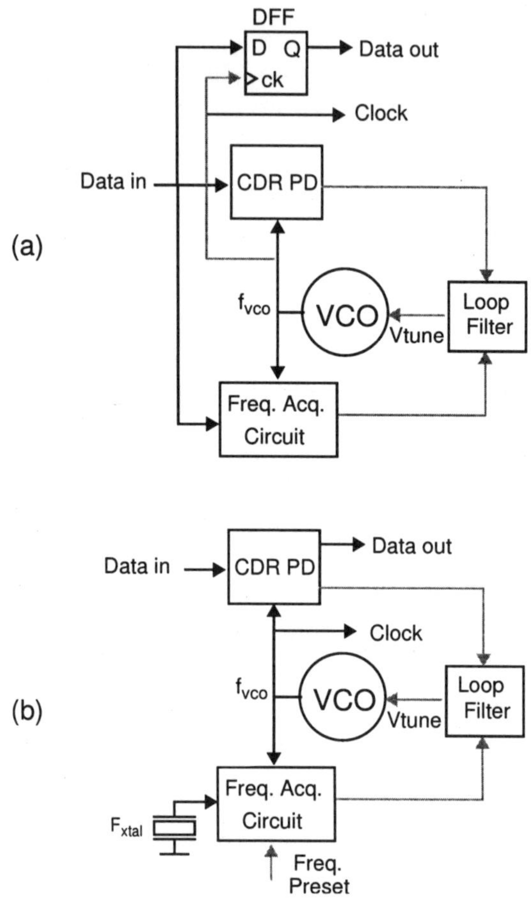

Fig. 1. Basic CDR architectures: (a) non-self correcting architecture, with a quadricorrelator as frequency acquisition aid. (b) self-correcting architecture, with frequency acquisition aid on a crystal oscillator.

In this paper, we will concentrate on a parallel CDR architecture which leads to higher operation speeds and low-power dissipation, with a comparable circuit complexity as the Alexander architecture. Linear CDR architectures for multiple octave operating range systems are also discussed. In addition, this paper covers a frequency acquisition architecture based on standard PLL building blocks, and key CDR building blocks, namely, multi-phase and wide tuning range oscillators.

2. Basic Concepts

A CDR is basically composed of five functional blocks, as depicted in Fig. 1(a): the CDR phase-detector (PD), the voltage-controlled oscillator (VCO), the loop filter, the frequency acquisition circuitry, and the D-type flip-flop (DFF), which is used for data regeneration. A programmable divider can be added in between the VCO and the PD in wide operating range systems.

Upon start-up, there may be a large frequency difference between the data signal frequency $f_b = 1/T_b$ (where T_b is a bit period) and the clock signal f_{vco}. The frequency acquisition circuit detects the frequency difference and generates a correction signal to bring the VCO frequency within the capture range of the phase detector. Note that the acquisition circuit of Fig. 1(a) extracts the frequency error information from the data, whereas the structure of (b) relies on a crystal reference. The pro's and con's of each approach will be discussed later.

As the VCO frequency enters a pre-defined frequency window centered on the data signal frequency, the frequency acquisition circuit turns itself off. The phase detector then aligns the phase of the VCO to the phase of the incoming data signal. The data signal is finally regenerated in the DFF, which responds to the rising edge of the clock (VCO) signal. For optimal operation of the CDR function, the active edge of the VCO should arrive at the moment the data signal presents a well defined level to the DFF, i.e., at the middle of the data-eye.

The system depicted in Fig. 1(a) is not a *self-correcting* architecture. The PD aligns the phase of the clock to the phase of the data at its input. The data, however, is regenerated in the DFF, a different circuit. Delay variations within the DFF itself, or on the path from the VCO output to the input of the DFF, are not corrected by the phase detector. The sampling moment at the DFF may shift from its optimum position at the middle of the data eye, and in a worse-case situation the DFF may sample the data at the transition moment. Therefore, reliability is likely to be a problem with the architecture of Fig. 1(a).

Self-correcting CDR architectures use phase-detector implementations, which use the same circuit elements to simultaneously assess the phase difference *and* to regenerate the data signal. In this way, delay variations in the logic elements are sensed and compensated by the phase detector. Such an architecture is depicted schematically in Fig. 1(b). Self-correcting architectures were proposed by Alexander,[5] Hogge, [4] DeVito,[6] and by Savoj and Razavi.[7,8] A variation of the Alexander detector is proposed in this paper. We will treat the self-correcting structures in more detail below.

3. High-Frequency CDR Architectures

3.1. *The Alexander CDR architecture*

Fig. 2 depicts the block diagram of the "Alexander" architecture.[5] The basic idea is that three data samples B, T and A are taken in a given bit interval. The data is sampled in DFF1 and DFF2. DFF3 and Dlatch1 provide "timing alignment" of the B,T and A samples for the phase detector. The sample T is compared to samples B and A in the phase detector. If $T = B$ and $T \neq A$ then the clock transition is advanced with respect to a data transition, and the phase detector delivers a Dn signal to the loop filter to decrease the frequency and delay the phase of the clock signal. If $T = A$ and $T \neq B$ the clock signal must be advanced in phase, and the phase detector generates an Up signal in that case. If $B = A$ no transition has occurred and no corrective action is carried out.

The Alexander architecture enables assessment of the data signal quality, by providing

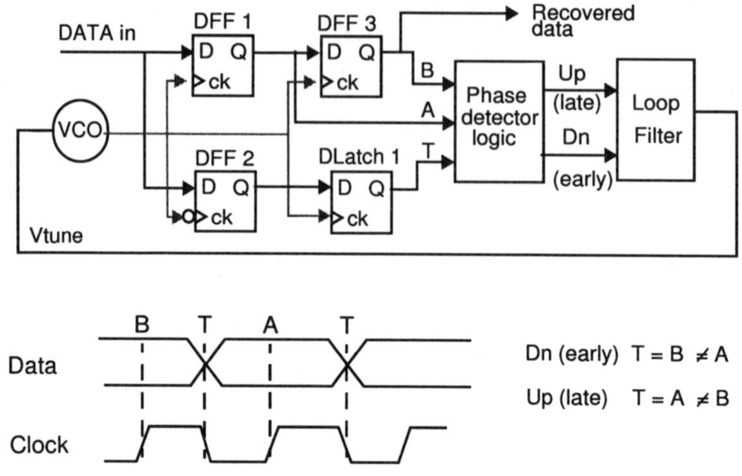

Fig. 2. Alexander CDR architecture.

a rough estimation of the bit-error-rate (BER). The BER assessment can be done by comparing the value of the A, B and T samples when the condition A equals B is fulfilled. In that case no transition has taken place during the previous interval, and the T signal is expected to be the same as A and B. If that is not the case then there must have been a "wrong" decision in one (or several) of the D flip-flops. Decision errors can only be explained by the presence of noise on the input data signal, so that a degradation in the quality of the transmission channel must have occurred. This property enables BER related loss-of-signal (LOS) detection.

The Alexander CDR architecture requires no analog pre-processing of the input data signal, enabling reliable operation over a wide range of input data-rates. In addition, it is self-correcting for delay variations, which is an important property for reliable high-speed operation. On the other hand, the flip-flops operate at the full data-rate frequency f_b, which leads to high power dissipation and to an unnecessary limitation of the maximum operation frequency in a given technology. In addition, the Up and Dn signals do not have a linear relationship to the phase error of the data and clock signals. This leads to difficulties in the prediction of the loop bandwidth and noise performance of the PLL.[9,10]

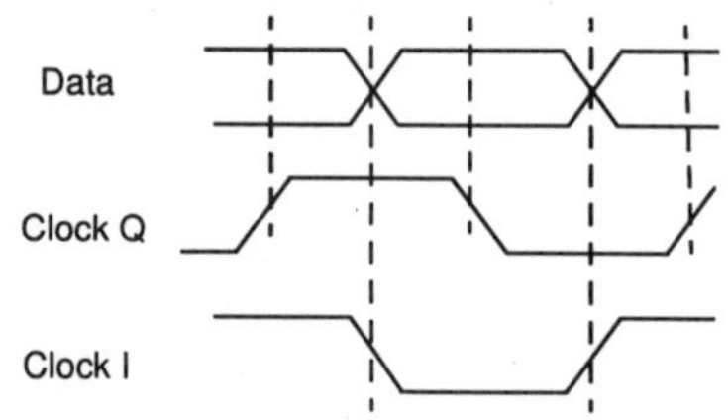

Fig. 3. Data and quadrature clock signals, in locked condition.

3.2. *Parallel CDR architecture*

The clock frequency of the Alexander CDR can be halved if the single-phase clock signal is replaced by clock signals with quadrature phase relationship CKI and CKQ. A synchronized situation is depicted graphically in Fig. 3. Transitions on the data occur (ideally) at the rising and/or falling edges of the CKI signal. As the edges of the CKQ signal are delayed by a quarter of a period with respect to transitions in CKI, they can be used to sample the data signal in the middle of the eye-diagram.

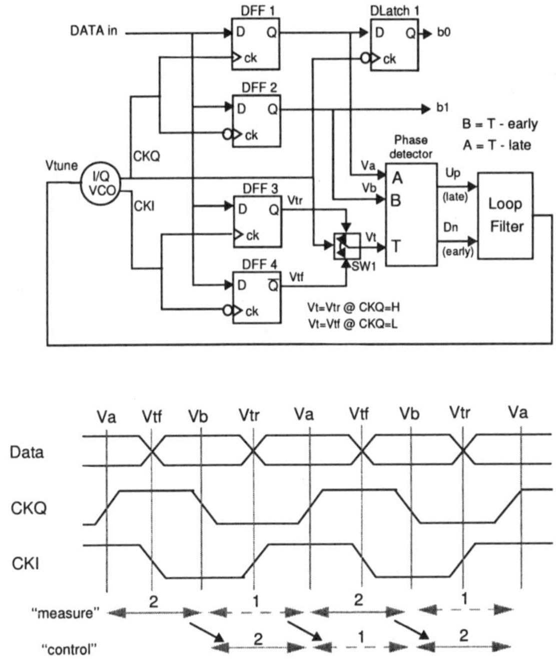

Fig. 4. Parallel CDR architecture with 1/2 rate clock and low complexity.

The parallel architectures described in the literature have twice as many logic elements as the Alexander architecture of Fig. 2.[11,12,13] The speed advantage is thus gained at the expense of doubled complexity, which leads to higher power dissipation and to a higher load on the clock buffer. We propose a parallel architecture with roughly the same circuit complexity as the original Alexander architecture, yet operating at one-half of the original clock frequency.[14] The block diagram is presented in Fig. 4.

The data stream is sampled at the rising edge of CKQ in DFF1 and at the falling edge of CKQ in DFF2. Similar operations are performed at the falling and rising edges of CKI, this time in DFF3 and DFF4. The samples obtained at the rising (V_a) and falling (V_b) edges of CKQ are compared to the samples obtained at the rising (V_{tr}) and falling (V_{tf}) edges of CKI. Note that Dlatch1 provides timing alignment of the demultiplexed data bits b_0 and b_1, and is not essential for operation of the CDR. Correction pulses are derived from the 1/2 rate samples V_a and V_b, and do not require an additional re-multiplexing action as in the

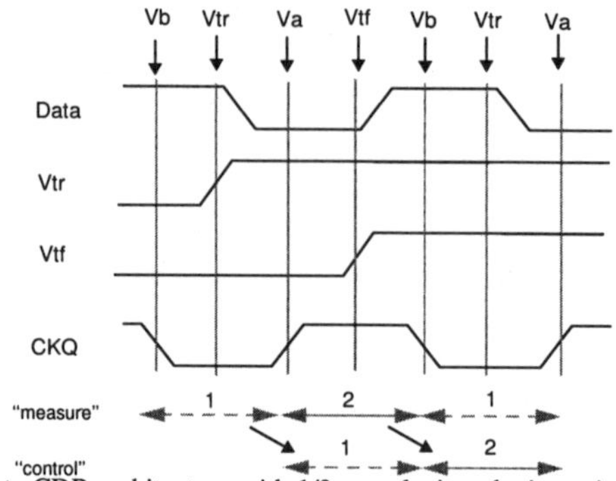

Fig. 5. The clue for the 1/2 rate CDR architecture with 1/2 complexity: the inversion of the data contained in the V_{tf} signal.

CDR architecture proposed by Savoj and Razavi.[8] In addition, the phase detector of Fig. 4 provides a tristate output, a convenient feature for systems in which long sequences of 0's or 1's may occur in the data.

With the architecture of Fig. 4, there is a "cyclical" and continuous phase comparison of three of the four available samples V_a, V_b, V_{tr} and V_{tf}. We define "measure" and "control" phases 1 and 2: a logic level CKQ= H corresponds to "control phase 1" and "measure phase 2"; CKQ= L corresponds to "control phase 2" and "measure phase 1". During "control phase 1", V_b is the "first" sample, V_{tr} is the transition sample and V_a is the "last" sample. In "control phase 2" the situation is reversed, V_a is the "first" sample, V_{tf} is the transition sample and V_b is the "last" sample.

Note the multiplexer in front of the T input of the phase detector in Fig. 4, and the sample V_{tf} which is the *inverted* data signal at the sampling moment. This inversion is the clue to correct functionality with reduced complexity. Fig. 5 shows a situation where the clock is advanced with respect to data transitions. Thus the phase detector must detect an "Early" condition and generate Dn signals. In "control phase 1" V_{tr} is connected to the T input of the detector, one has $B = T$, and the correct timing information is generated.

During "measure phase 2" the data is low during the sampling moments of V_a and V_{tf}, and high at the moment the next sample for V_b is obtained; however, due to the inversion, V_{tf} is high during "control phase 2". This leads to the same situation as during "control phase 1", $B = T$, and the "Early" condition is correctly detected during "control phase 2" as well.

4. Linear CDR Architectures

This section focuses on linear phase detectors for PLL-based CDR systems with multiple octave operating range, suitable for multi-gigabit transmission systems.[15] Such a CDR supports low bitrate legacy systems already installed in the field, as well as high-bitrate systems of today and potential new standards of the (near) future, originating in other application areas than traditional data transmission markets.

The multiple octave capability of the CDR rules out any circuitry based on delay,[16]

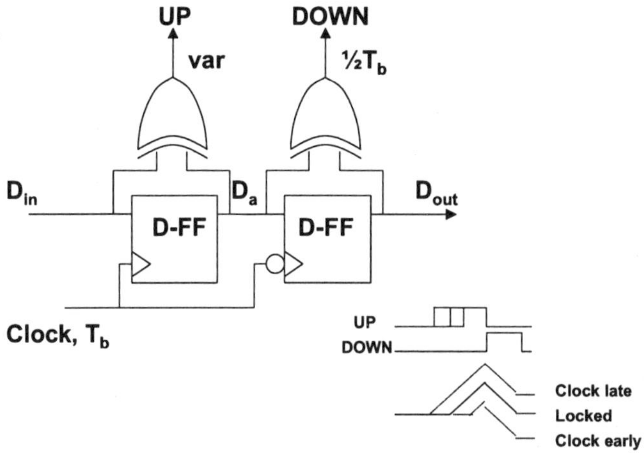

Fig. 6. Hogge's linear self-correcting phase detector.

simply because the bitrate variation is to wide to cover, for instance from 30 Mbit/s up to 3.2 Gbit/s. Over this enormous span the performance of the CDR has to be "the same" when expressed in relation to the bitrate, i.e., the absolute performance changes with the bitrate, but when expressed in unit interval (UI) the performance must be constant. This requires all parameters and characteristics of the CDR to be relative to the bitrate.

4.1. *Phase detector*

In an any-rate system the phase detector should be "self-correcting". Due to its quantized output, a PLL employing an Alexander phase detector is known as a "bang-bang" loop. At high frequencies a practical implementation of the phase detector behaves more linear, due to circuit limitations. The detectors transfer characteristic degrades from a hard limiter at low-bit rates into a more linear hyperbolic tangent at high bit-rates. So the phase detection characteristic is not constant over multiple octaves and hence the CDR's loop behavior will be different per octave, not complying to the requirement of the previous paragraph. The efficient use of hardware in the parallel architecture of Fig. 4 makes it more suitable for high frequency operation, however it has the same drawback.

More predictable behavior can be achieved with linear phase detectors, where the correction signal depends linearly on the phase error. Examples of self correcting linear phase detectors, suitable for phase error extraction from random data, are Hogge's phase detector,[4] the Triwave phase detector,[6] and the phase detector proposed by Savoj and Razavi.[7]

These detectors are based on the same principle, which can be explained with reference to the Hogge detector depicted in Fig. 6. The incoming data is sampled with a D flip-flop. An exclusive-OR gate (XOR) compares the instantaneous input data D_{in} with the data sampled one clock cycle T_b before (D_a). The XOR output will be high from the moment a data transition occurs till the next rising clock edge. The duration of the "up" signal is thus

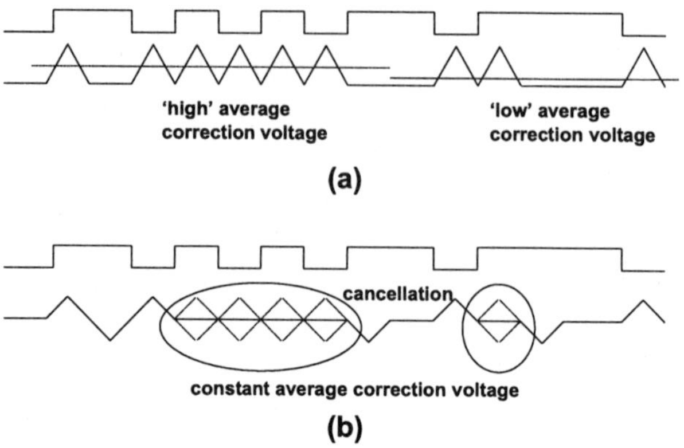

Fig. 7. Correction signal waveforms;(a) Hogge's detector (b) Triwave detector.

a linear function of the arrival time of a data transition relative to the rising clock edge. A second D flip-flop and XOR gate are used to derive a correction signal with a fixed width of half a clock cycle ($0.5T_b$), the "down" signal. The net difference in duration of the "up" and "down" signal represents the phase error between the data transition and the rising clock edge.

The phase error can only be established by integration over one clock period, since the "up" and "down" signal do not coincide. Even if there is no phase error (the "up" and "down" signal have equal width), the integrated correction signal will be triangular. Although the integrated signal will have the same value at the end of the clock interval as at the beginning, representing the same frequency and phase, the DC content of the signal depends on the data density. When many data transitions occur in a certain time interval, as in Fig. 7a, the DC content of the correction triangles will be high. When data transitions are sparse, the DC content of the correction signal will be much lower. This data density dependency is due to the uni-polar nature of the correction signal (positive triangles only), and it results in pattern dependent jitter.

The Triwave detector, depicted in Fig. 8, circumvents this problem by virtue of a bi-polar correction signal (combined positive and negative triangles). One data transition invokes four correction signals, one with variable duration and three with fixed width. The integrated correction signal has no DC content and is hence data density independent. The phase detection cycle takes two clock periods, causing two phase comparison cycles to overlap when single bits occur in the data stream. In these cases the correction signals of the rising data transition and the falling data transition will partly cancel (see Fig. 7b). Hence the triwave correction signal still shows some data dependency.

Savoj's detector, depicted in Fig. 9, employs Hogge's principle in parallel form, halving the required clock frequency and demultiplexing the data stream simultaneously. One of the input latches in Fig. 9 is transparent, representing the instantaneous input data and the other is latched, storing the previous data. The first XOR gate compares the output signals of both lathes, producing an "up" signal. Since both signals have propagated through the latches, no delay adjustment is necessary to achieve zero phase error at equilibrium. The

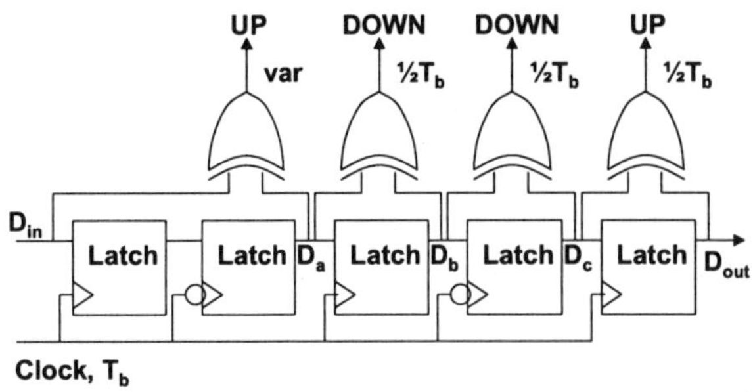

Fig. 8. Triwave phase detector.

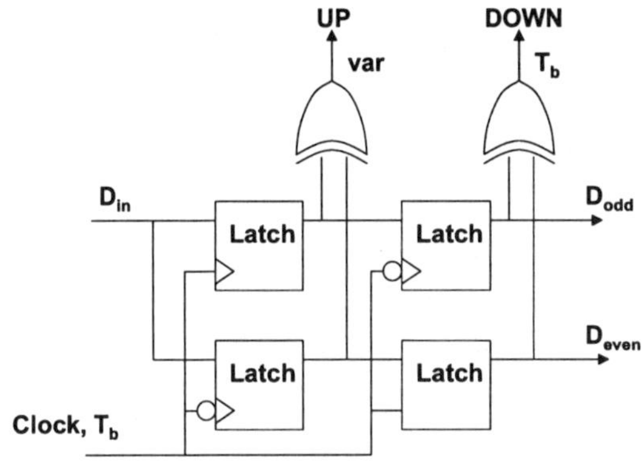

Fig. 9. Savoj's parallel implementation of a linear phase detector.

second set of latches and XOR gate again derive a "down" correction signal.

At zero phase error the "down" signal is twice as long as the "up" signal, so the current associated with the "down" signal must be halved. At high bit rates the shape of the "up" and "down" pulse can be quite different, resulting in an undesired phase offset. Sampling point optimization can be achieved in all linear detectors by changing the ratio of up current to down current. The imbalance causes an offset that can only be corrected with a phase error, shifting the loop equilibrium to another position in the eye diagram.

5. Frequency Acquisition Aid Architectures

5.1. *Quadricorrelators*

In the architecture of Fig. 1(a) the frequency acquisition circuitry will most likely be an implementation of a *quadricorrelator* circuit.[17-21] The main draw-back of the quadricorrelator is that it relies on transitions of the data signal, in order to assess the frequency difference between the data and the VCO signal. Therefore, quadricorrelators are not well suited for use in SONET/SDH applications, where long sequences of 0's or 1's may occur in the data. On the other hand, data communication in e.g., computer networks normally use data coding (such as the 8B/10B code)[19] to ensure a minimum number of transitions

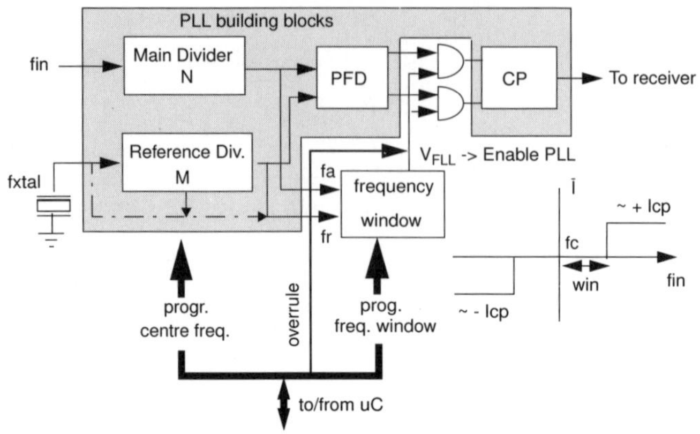

Fig. 10. Block diagram depicting the combined PLL/FLL tuning system.

per time-interval and a DC-free data content. However, the best case frequency error detection range is in the order of ±25%.[6,17,20] The suitability of the quadricorrelator as the frequency acquisition circuitry is thus strongly linked to the intended application. Its main advantage is that it requires no external reference frequency.

5.2. *Combined PLL/FLL architecture*

The architecture of Fig. 1(b) uses a crystal oscillator as a reference source for the frequency acquisition circuit. In the frequency acquisition phase the system operates as a standard frequency synthesizer, so that the oscillator frequency is brought within a predetermined (programmable) frequency error from the data signal frequency f_b, even in the absence of data transitions. By means of the frequency preset input, the frequency of the oscillator may be programmed to "any" frequency, so that multi-rate systems can be easily implemented with this concept.

An architecture based on a PLL frequency synthesizer (PLL-FS) is the natural choice for the frequency acquisition circuitry depicted in Fig. 1(b), as they are built with well-known and well-optimized building blocks.[22] However, a standard PLL-FS is not suited for synchronising the VCO of an CDR function, as two PLLs would try to phase-lock the same VCO to different signals, one to the multiplied crystal and the other to the data signal.

We propose a combined PLL/FLL tuning system as depicted in Fig. 10.[23] The main difference is the addition of the frequency window block. The blocks in the shaded area are original building blocks from PLL synthesizers, which can be used in this concept without architectural modifications. The (programmable) frequency window block generates a range of frequencies around the data signal frequency in which the PLL-FS outputs no signal to the loop filter.

When the input frequency f_{in} is far away from the desired tuning frequency f_c, the frequency window block delivers an *Enable* signal V_{FLL} for the PFD/CP, and the system then operates as a normal PLL-FS. The frequency window block activates/deactivates the PLL action automatically, as the frequency f_{in} comes inside or goes outside of the frequency

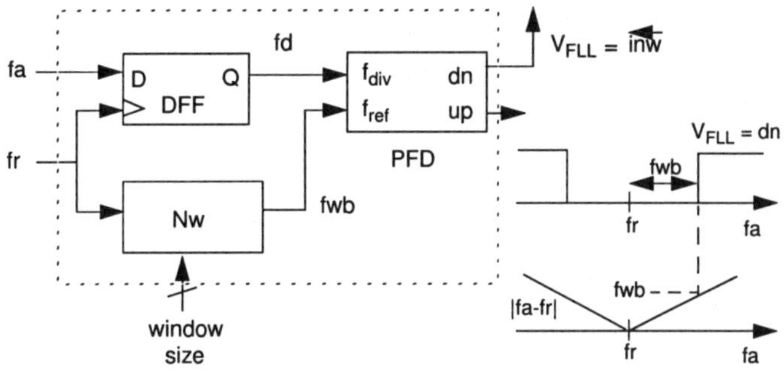

Fig. 11. Possible implementation of the frequency window detector function.

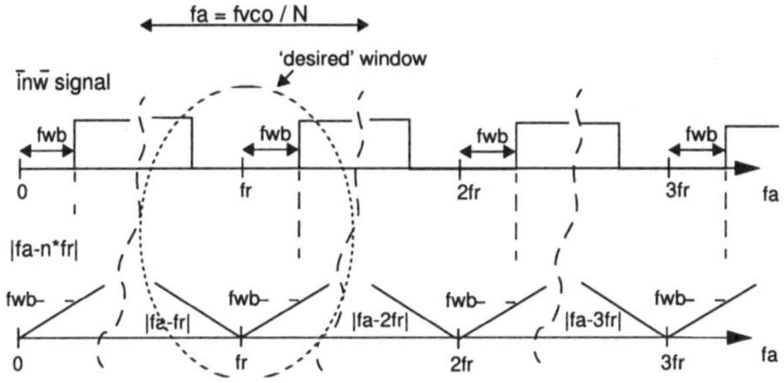

Fig. 12. "Desired" and "undesired" frequency window detection, due to sampling process in the D flip-flop.

window $f_c \pm win$, with win as depicted in Fig. 10.

A possible implementation of the frequency window block is depicted in Fig. 11. It comprises a D type flip-flop DFF, a programmable divider Nw and a phase frequency detector PFD. The divider Nw and the PFD are similar to the PLL building blocks shown in the shaded area of Fig. 10.

Remember that the PFD is able to detect frequency differences at its inputs, and to signal which of the two input signals has the larger frequency. If the frequency at the input f_{div} is larger than that at the input f_{ref}, the signal at the output node dn becomes active, though not necessarily with 100% duty cycle. The D flip-flop samples the signal f_a with sampling frequency f_r. If f_a has 50% duty cycle and its frequency is larger than $f_r/2$, the output f_d is an aliased version of the signal f_a. In other words, f_d has a frequency equal to the modulus of frequency difference $|f_a - n \times f_r|$, with $n = 1, 2, \ldots$

Consider the case were $n = 1$ $(0.5 \times f_r < f_a < 1.5 \times f_r)$. The frequency window block informs whether the frequency of the f_d signal ($|f_a - f_r|$ in this case) is larger or smaller than the frequency of the signal f_{wb}. If it is larger, then dn becomes active, signalling an out-of-window situation. If it is smaller, dn stays low, signalling an in-window situation. A programmable frequency window size can be realized by simply changing the division ratio Nw.

Fig. 12 shows that undesired frequency windows are present at frequencies in the neighborhood of higher harmonics of $f_r (n = 2, 3, ...)$, in addition to the one close to zero Hz. It is important to be aware of this fact, in order to take measures (if necessary) to avoid the PLL part of the combined tuning system being disabled before the correct operation frequency has been reached. An easy way to solve this issue is to overrule the *Enable PLL* signal for a certain time, until the VCO is for sure in the neighborhood of the desired tuning frequency ($f_a \sim f_r$).

6. CDR Building Blocks

6.1. *Wide-tuning range oscillator for multi-rate applications*

In order to achieve multiple octave capability in any-rate CDR architectures, the oscillator must at least be tunable over an octave under all conditions; process, supply voltage and temperature. Lower octaves will be supported by dividing the oscillator's frequency by 2, 4, 8 etc. The oscillator not only has to be tunable over at least an octave, but also has to maintain its properties over this range, like amplitude, phase noise and tuning constant (K_{vco}). Especially the latter one is important for achieving predictable loop behavior and 'constant' (relative) loop bandwidth over the octave(s).

A bipolar implementation of an oscillator fulfilling these requirements is shown in Fig. 13. The oscillator is a two-stage ring oscillator. The unit cell consists of 2 amplifiers, one fast and the other slow. The extra base resistors form a current dependent delay with the input capacitance of the second differential pair. The output waveform of the unit cell is the sum of the fast and the slow branch. By varying the amount of current in the fast and the slow branch, the net delay from input to output of the cell can be controlled. Note that the sum of the fast and the slow current is constant, yielding constant amplitude over the entire tuning range. Proper selection of the delay caused by R_{s1}/R_{s2} and the input capacitances of Q_{s1}/Q_{s2} (big transistors) yield almost linear tuning over more than an octave (see Fig. 14). This topology also maintains consistent phase noise performance over its entire tuning range.

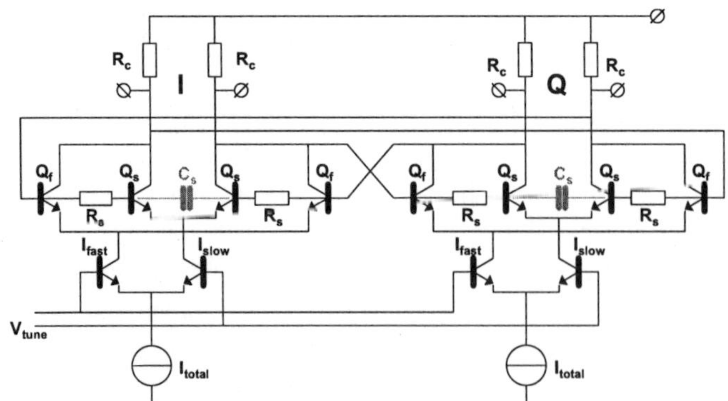

Fig. 13. Ring oscillator with wide tuning range.

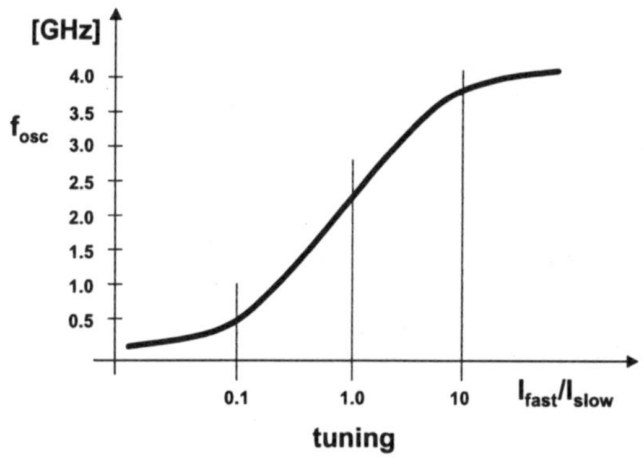

Fig. 14. Ring oscillator tuning characteristic.

6.2. *I/Q LC oscillators for high-frequency CDR architectures*

Many advanced CDR architectures require multi-phase oscillators. For example, a 40 Gb/s CDR architecture employing a quarter-rate phase detector uses a four-stage differential multi-phase oscillator.[24] Similarly, I/Q oscillators are needed in half-rate CDR architectures, which operate at half the frequency of the incoming data rate, see Fig. 4. Blind oversampling CDRs perform phase detection and data recovery purely in the digital domain, and need a multi-phase oscillator to collect the samples for the data recovery algorithm they employ.[25]

Multi-phase signals are easily generated with (inductor-less) ring oscillators. Other advantages of ring oscillators are their compactness in terms of chip area compared to LC oscillators, and their relatively large tuning range. However, because of the energy-preserving character of resonators (with a quality factor Q>1), it is well known that LC oscillators exhibit much better phase noise performance for a given power budget.[26] Also, especially for high-frequencies (e.g. > 1/4 of the transition frequency f_T of the implementation technology), it is increasingly difficult to achieve acceptable phase noise performance with ring oscillators.[27] At low frequencies (relative to the f_T), all transistors in the signal-path of high-performance ring oscillators switch, effectively reducing the phase noise (this can be modeled by a higher effective quality factor).[28] At high frequencies, switching waveforms become sinusoidal, and phase noise increases. In summary, when phase noise requirements or the power budget are tight, and multi-phase high-frequency signals are required in a CDR architecture, a multi-phase LC oscillator is the right choice.

6.3. *Multi-phase LC oscillator properties*

A behavioral model of a multi-phase LC oscillator is shown in Fig. 15.[29,30]

Fig. 15 shows how N LC oscillators with negative transconductance gm_l that are coupled with transconductances gm_c, form a multi-phase oscillator. Expressed in terms of resonator phase shift ϕ_{res} and quality factor Q_p, and resonance frequency ω_{LC} of the resonator of one stage in Fig. 15, the oscillation frequency of a N-stage LC oscillator can be

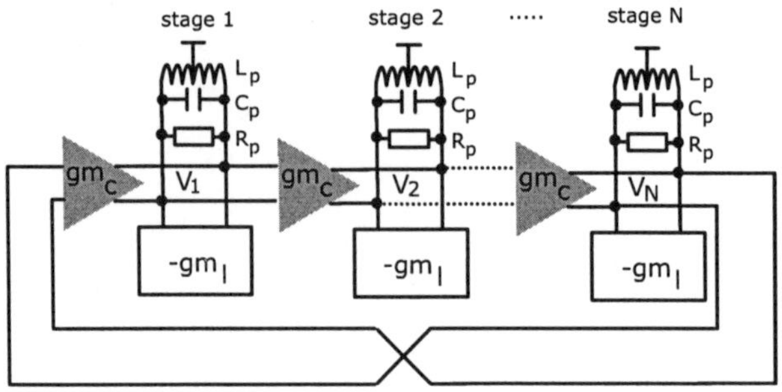

Fig. 15. A behavioral model of N LC oscillators coupled with transconductances gm_C.

written as[30]

$$\omega_{NLC} = \frac{-\tan(\phi_{res}) + \sqrt{4\,Q_p^2 + \tan^2(\phi_{res})}}{2Q_p} \cdot \omega_{LC}. \tag{1}$$

Equation (1) shows that for $\phi_{res} = 0°$ the frequency ω_{NLC} of the multi-phase LC oscillator becomes equal to that of a single-phase LC oscillator. The oscillator can be tuned by varying ϕ_{res} (phase-shift tuning) or by varying the self-resonance frequency ω_{LC} of the resonator in each single stage.

The effective quality factor of an N-stage LC oscillator can be approximated by[31]

$$Q_{NLC} \approx N \cdot Q_p \cdot \cos(\phi_{res}), \tag{2}$$

which reduces to NQ_p for zero resonator phase shift. As argued in the literature,[30] the quality factor of a multi-phase oscillator can be maximized ($\phi_{res} = 0$, leading to best phase noise performance) by implementing transconductances gm_c from Fig. 15 such that they exhibit a phase shift of $180°/N$. Alternatively, other coupling methods can be employed for optimum phase noise performance, such as common-mode inductive coupling in an I/Q LC oscillator.[32]

6.4. *A 5 GHz phase-shift tuned I/Q LC oscillator*

Two-stage I/Q LC oscillators are optimally coupled if each stage is connected with a phase shift of ±90°. An example of an I/Q oscillator with phase shifters is shown in Fig. 16. The circuit is based on a design presented before,[31] with the resonator layout reused to save design time. The desired 90° phase shift is implemented by the parasitic phase shift of the coupling transistors (with tail current I_{couple}) and the emitter followers (with bias current I_{buf}). The cross-coupled pair implements a negative resistance and is used to set (with tail current I_{level}) the voltage swing across the resonator. The tuning behavior predicted by (1) can be verified for different Q_p values with an integrated method to vary the resonator quality factor. Variation of Q_p is taken care of by the varactors shown in Fig. 16, making use of the varying series resistance as function of reverse bias of PN-junction type varactors. Simulated Q_p at 5 GHz is ≈ 3 for V_{tune} = 2.7 V and ≈ 8 for V_{tune} = 0 V.

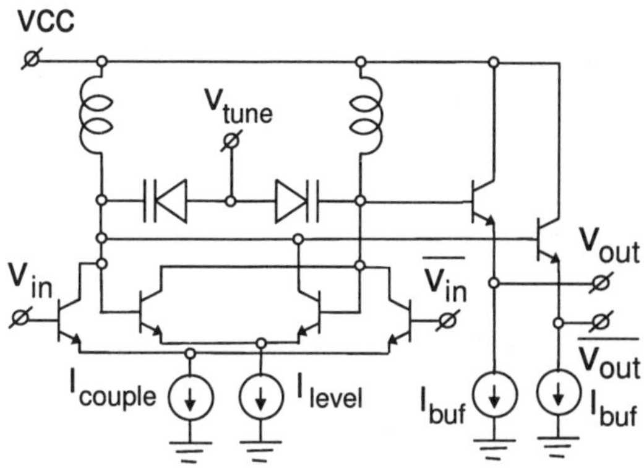

Fig. 16. Circuit diagram of one stage of the I/Q LC oscillator.

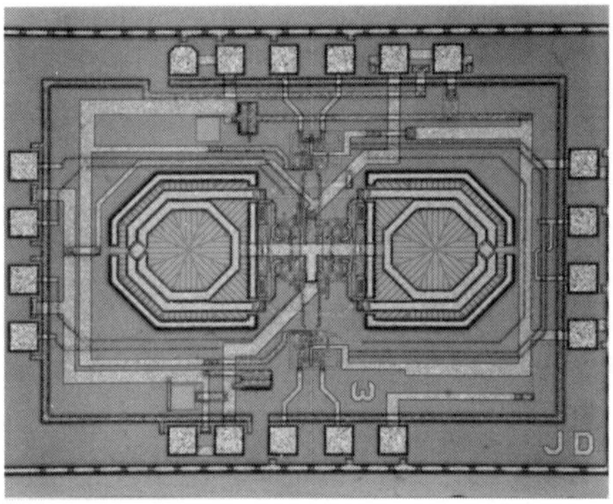

Fig. 17. Micro-graph of the I/Q LC oscillator.

6.5. *Experimental results*

Fig. 17 shows the micro-graph of the realized I/Q ring oscillator with a center frequency of around 5.3 GHz.[33] The active chip area is $1450\mu\text{m} \cdot 2280\mu\text{m}$. The power dissipation of the VCO (excluding 50 ohm buffers) is 45 mW (at 5.3 GHz) with a 2.7 V supply voltage.

The measured tuning characteristics versus $I_{phase-shifter} = I_{couple} + I_{buf}$ for $Q_p = 3$ and $Q_p = 8$ are shown in Fig. 18. Because the quality factor of the resonator Q_p is a function of the varactor voltage the slope of the tuning curves versus $I_{phase-shifter}$ becomes steeper for lower values of Q_p, as predicted by (1). Around $I_{phase-shifter} = 6$ mA, the curves cross and ω_{NLC} equals ω_{LC} (the factor in front of ω_{LC} in (1) reduces to unity for $\phi_{res} = 0$). For high values of current $I_{phase-shifter}$, the phase shift of the buffers and coupling transistors are no longer a function of their biasing current and the tuning characteristics saturate.

Worst case measured phase noise is -102 dBc/Hz at 2 MHz offset of a 4.7 GHz carrier. At this frequency the carrier is only a few 100 mV as a result of a minimum $I_{phase-shifter}$. At 5.6 GHz the measured $\mathcal{L}$(2MHz) is better than -108 dBc/Hz. The I/Q oscillator meets the phase noise requirements for (half-rate) CDR circuits for the SONET/SDH 10 Gb/s

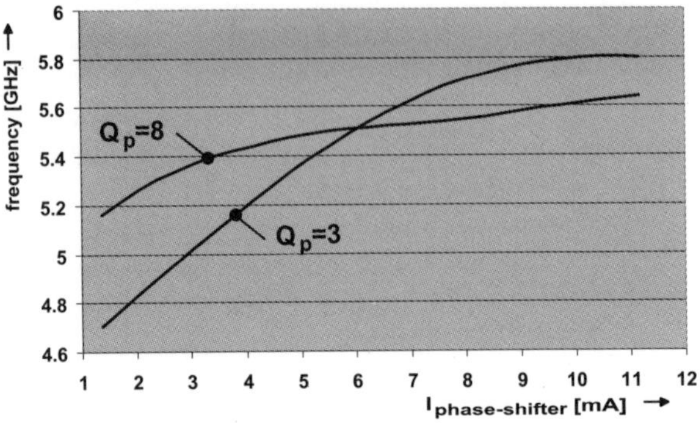

Fig. 18. Measured oscillation frequency versus $I_{phase-shifter}$ for $Q_P = 3$ and $Q_P = 8$.

standard (OC-192/STM-64).[27] Using phase-shift tuning, the tuning range is large enough to cover process spread.

7. Conclusions

We focused on high performance CDR architectures and building blocks. A parallel CDR architecture for high operation speeds and low-power dissipation was described, followed by a linear CDR architecture with multiple octave operating range. A frequency acquisition architecture, based on standard PLL building blocks, was introduced, before discussed important building blocks, namely multi-phase and wide tuning range oscillators.

8. References

1. M. Sexton and A.Reid, *Broadband Networking: ATM, SDH and SONET*, Norwood, MA: Artech House, 1997
2. B. Razavi (Ed.), *Monolithic Phase-locked Loops and Clock Recovery Circuits*, Piscataway, NJ: IEEE Press, 1996
3. B. Razavi (Ed.), *Phase-Locking in High-Performance Systems*, Piscataway, NJ: IEEE Press, 2003
4. C.R. Hogge, "A self correcting clock recovery circuit", *IEEE Journal of Lightwave Technology*, vol. LT-3, pp. 1312-1314, Dec. 1985.
5. J.D.H. Alexander, "Clock recovery from random binary signals", *Electronic Letters*, vol. 11, pp. 541-542, Dec. 1975.
6. L.M. DeVito, "A versatile clock recovery architecture and monolithic implementation", In [2], pp. 405-420, 1996.
7. J. Savoj and B. Razavi, "A 10-Gb/s CMOS clock and data recovery circuit with a half-rate linear phase detector", *IEEE Journal of Solid State Circuits*, vol. 36, pp. 761-768, May 2001.
8. J. Savoj and B. Razavi, "A 10-Gb/s CMOS clock and data recovery circuit with a half-rate binary phase/frequency detector," *IEEE Journal of Solid State Circuits*, Vol. 38, pp. 13–21, Jan. 2003
9. J. Lee, K.S. Kundert and B. Razavi, "Modelling of jitter in bang-bang clock and data recovery circuits," *IEEE Custom Integrated Circuits Conference*, pp. 711–714, 2003
10. Y. Choi et al., "Jitter transfer analysis of tracked oversampling techniques for multigigabit clock and data recovery," *IEEE Transactions on Circuits and Systems:II*, vol. 50, pp. 775–783, Nov.

2003

11. A. Buchwald and K. Martin, *Integrated Fiber-Optic Receivers*, Boston, MA: Kluwer Academic Publishers, 1995
12. J. Hauenschild et al., "A plastic packaged 10Gb/s BiCMOS Clock and Data Recovering 1:4-demultiplexer with external VCO", *IEEE Journal of Solid State Circuits*, vol. 31, no. 12, pp. 2056-2059, Dec. 1996.
13. M. Reinhold et al., A fully integrated 40-Gb/s clock and data recovery/1 : 4 DEMUX IC in SiGe technology, *IEEE Journal of Solid State Circuits*, vol. 36, pp. 1937-1945, Dec. 2001.
14. C.S. Vaucher, "Data clock recovery circuit", European Patent EP1183781, June 2002.
15. Philips Semiconductors, *TZA3015HW 30 Mbit/s to 3.2 Gbit/s A-rate 4-bit fibre optic transceiver*, Datasheet, Dec. 2003
16. T.H. Lee and J.F. Bulzachelli, "A 155 MHz clock recovery delay- and phase-locked loop," *IEEE Journal of Solid State Circuits*, vol. SC-27, pp. 1736–1746, Dec. 1992.
17. J.A. Bellisio, "A new phase-locked timing recovery method for digital regenerators", *IEEE Int. Comm. Conf. Rec.*, pp. 10.17–10.20, June 1976.
18. D.G. Messerschmitt, "Frequency detectors for PLL acquisition in timing and carrier recovery", *IEEE Transactions on Communications*, vol. COM-27, pp. 1288–1295, Sept. 1979.
19. M. Soyuer, "A monolithic 2.3-Gb/s 100mW clock and data recovery circuit in silicon bipolar technology", *IEEE Journal of Solid State Circuits*, pp. 1310 – 1313, Dec. 1993
20. A. Pottbaecker and U. Langmann, "An 8 GHz silicon bipolar clock recovery and data regenerator IC", *IEEE Journal of Solid State Circuits*, pp. 1572 – 1576, Dec. 1994.
21. R. van de Beek et al., "A 2.5 to 10GHz clock multiplier unit with 0.22ps RMS jitter in a 0.18 μm CMOS technology," *IEEE Int. Solid-State Circuits Conference*, pp. 178–179, 2003
22. C.S. Vaucher, *Architectures for RF Frequency Synthesizers*. Boston, MA: Kluwer Academic Publishers, 2002
23. C.S. Vaucher and D. Kasperkovitz, "Frequency synthesizer with a digital frequency window detector and a receiver incorporating the same", American Patent US6366769, April 2002.
24. J. Lee and B. Razavi, "A 40-Gb/s clock and data recovery circuit in 0.18μm CMOS technology", *IEEE Journal of Solid State Circuits*, vol. 38, no. 12, pp. 2181–2190, Dec. 2003.
25. J. Kim and D. Jeong, "Multi-gigabit-rate clock and data recovery based on blind oversampling", *IEEE Communications Magazine*, pp. 68–74, Dec. 2003.
26. D. Leenaerts, J. van der Tang, and C. Vaucher, *Circuit Design for RF Transceivers*, Boston, MA: Kluwer Academic Publishers, 2001.
27. J. van der Tang, D. Kasperkovitz and A. van Roermund, "A 9.8-11.5 GHz quadrature ring oscillator for optical receivers", *IEEE Journal of Solid State Circuits*, vol. 37, no. 3, pp. 438-442, Mar. 2002.
28. L. Dai and R. Harjani, *Design of high-performance CMOS voltage controlled oscillators*, Boston, MA: Kluwer Academic Publishers, 2002.
29. J. Kim and B. Kim, "A low-phase-noise CMOS LC oscillator with a ring structure", *IEEE Int. Solid-State Circuits Conference*, pp. 430–431, 2000.
30. J. van der Tang, D. Kasperkovitz and A. van Roermund, *High-Frequency Oscillator Design for Integrated Transceivers*, Boston, MA: Kluwer Academic Publishers, 2002.
31. J. van der Tang et al., "Analysis and design of an optimally coupled 5 GHz quadrature LC oscillator", *IEEE Journal of Solid State Circuits*, vol. 37, no. 5, pp. 657–661, May 2002.
32. S. Gierkink et al., "A low-phase-noise 5 GHz quadrature CMOS VCO using common-mode inductive coupling", *ESSCIRC*, pp. 539–542, 2002.
33. J. van der Tang and A. van Roermund, "A 5.3 GHz phase-shift tuned I/Q LC oscillator with 1.1 GHz tuning range", *IEEE Radio Frequency Integrated Circuits Symposium*, pp. A133–A136, 2003.

International Journal of High Speed Electronics and Systems
Vol. 15, No. 3 (2005) 599–614

World Scientific
www.worldscientific.com

MOS CURRENT MODE LOGIC CIRCUITS: DESIGN CONSIDERATION IN HIGH-SPEED LOW-POWER APPLICATIONS AND ITS FUTURE TREND, A TUTORIAL

YUYU LIU

Department of Electronic Engineering, Tsinghua University
Beijing 100084, China
liu-yy02@mails.tsinghua.edu.cn

JINGUO QUAN

Tsinghua Shenzhen Graduate School, Xili Lake
Shenzhen 518000, China
quanjg@mail.sz.tsinghua.edu.cn

HUAZHONG YANG

Department of Electronic Engineering, Tsinghua University
Beijing 100084, China
yanghz@tsinghua.edu.cn

HUI WANG

Department of Electronic Engineering, Tsinghua University
Beijing 100084, China
wangh@tsinghua.edu.cn

In this paper, a logic style that is becoming increasingly popular is presented, which is called MOS Current Mode Logic (MCML). MCML is a novel and useful logic style for high-speed, low-power and mixed-signal applications. Its high-speed switching, low supply voltage and reduced output voltage swing contribute to its high performance, low power dissipation, and low noise features. MCML circuits are compared to several other logic styles, such as conventional static CMOS, dynamic logic, and traditional emitter coupled logic (ECL) in terms of power, delay and common mode noise immunity. MCML circuits seem to be very promising in high-speed, low-power and mixed-signal digital circuit applications, such as portable electronic devices, gigahertz microprocessors, and optical transceivers.

Keywords: MOS current mode logic; high-speed; low-power.

1. Introduction

The advances in VLSI technology have made portable electronic devices become more and more popular today. Laptop computers, cellular phones, and personal digital assistants have all become more prevalent in people's lives each year. Except for the high-speed demand, another most significant properties of portable devices are battery-powered and lightness. As CMOS technology scales down and device density increases, the power consumption of future IC's will increase excessively beyond the traditional constrains of packaging cost and heat dissipation. The short battery life and the extra weight of the batteries will become unbearable if significant ameliorations aiming at low-power are not made. On the other hand, as feature size of transistor scales down, it is extremely desirable to integrate analog and digital circuits onto the same die for many communication systems in order to reduce total system area and production costs. However, it is hard to realize this integration because of the difficulty in designing high precision analog circuits in the presence of large digital switching noise. The difficulties mentioned above will also increase as process technology scales down further. Therefore, breakthroughs in design of low-power, high-speed and high-density circuits become extremely imperative for future integrated circuits.

Low-power, high-speed, large integration density and low cost design have been the most significant goals in today's VLSI design. Quite a number of these requirements can be achieved by improving process technology, such as shrinking devices. However, the circuits in the future will require greater improvement than are obtained by improving process technology. The circuit design innovation to achieve the high-speed, low-power and high density goals by using present process technology is necessary for realization of future circuits.

This paper introduces a novel logic style, which is called MOS Current Mode Logic (MCML) style. Considerable potential for speed improving, power saving and noise reducing can be obtained by using MCML circuits for implementing digital logic circuits. MCML circuits have good features of both MOS devices (such as large yield, high density and low cost) and traditional current mode circuits (such as reduction in dI/dt effects, common mode noise immunity, process and voltage variation immunity and small signal voltage swing). MCML circuits have differential structure and can be used to implement any logic functions. Because MCML circuits are power-efficient at high operating frequency and output complementary signals, high-speed complex circuits (such as exclusive-or (EXOR) / exclusive-nor (EXNOR) gates) are more feasible to be constructed with MCML style.

This paper begins with an introduction of basic MCML gate configuration and operation. Then the performance of MCML compared with several other logic styles will be discussed and the appropriate domains of performance in which MCML presents benefits over other current logic styles will be evaluated. The next section analyses the special merits of MCML circuits and presents typical application examples. In order to utilize MCML circuits more efficiently, several improved MCML circuits will be briefly introduced. Finally, some conclusions are drawn and future work is preconceived.

2. Basic MCML Circuits

2.1. *Circuit configuration and operation*

The ideal configuration of a basic MCML gate is shown in Fig. 1.[1]

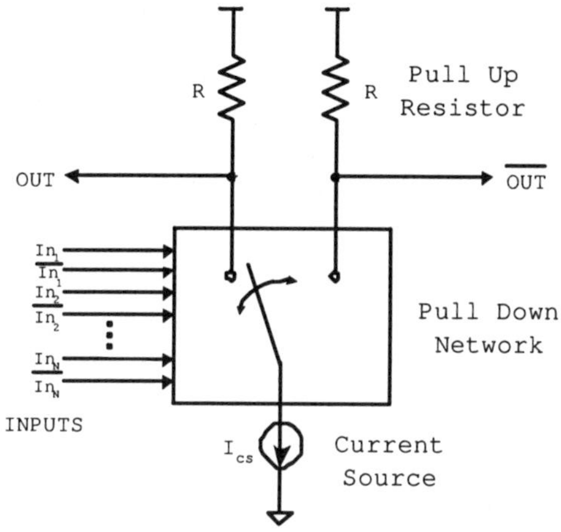

Fig. 1. Basic MCML gate structure.

An MCML gate mainly consists of three parts: two pull-up resistors, a pull-down network and a constant current source. The pull-down network, which is connected between the pull-up resistors and the current source, is fully differential and similar to other differential logic styles such as differential cascade voltage switch logic (or DCVSL). The pull-down network is constructed with sets of differential pairs. Every differential pair consists of two MOS transistors, which are common-source connected and controlled by two differential input signals respectively. Thus, every input signal as well as its complement must be available to an MCML gate.

The principle of operation of an MCML gate depends on current steering. The pull-down network steers current between the two pull-up resistors and its operation is based on the differential pair circuit. Every input signal and its complement control the current flowing through two transistor branches. The output voltage level is based on the difference between two currents passing through two pull-up resistors. The pull-down network should be designed so that, whatever the value of inputs, one and only one pull-up resistor is connected to the current source. That means one branch will implement the required function, while the other branch implements its inverse function.

For an MCML inverter shown in Fig. 2, the pull-down network is a differential pair, which is the simplest case for the pull-down network. The current source is implemented using an NMOS transistor controlled by a fixed gate voltage (Vref) working in the saturation region. Vref determines I_{cs}, the magnitude of current

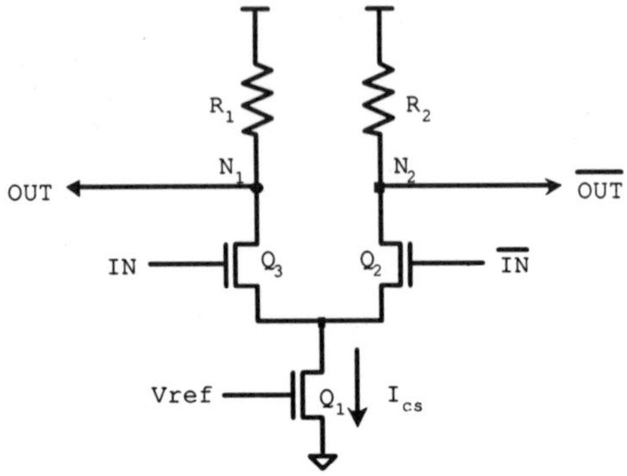

Fig. 2. MCML inverter.

of the current source. For example: if $V_{IN} > V_{\bar{IN}} \rightarrow V_{gs}(Q_3) > V_{gs}(Q_2) \rightarrow$ Q_3 is ON and saturation, Q_2 is OFF $\rightarrow I(Q_3) > I(Q_2), I(Q_3) \approx I_{cs}, I(Q_2) \approx 0 \rightarrow$ $V_{N1} = LOW \approx V_{dd} - I_{cs}R, V_{N2} = HIGH \approx V_{dd}$, where R is the value of a pull-up resistor, V_{dd} is the supply voltage, LOW represents low output voltage level and $HIGH$ represents high output voltage level. Output voltage swing (ΔV) of an MCML gate is defined as the voltage difference between two output nodes N_1 and N_2 at steady state. In the above example, $\Delta V = I_{cs}R$. The amount of current passing through the ON branch controls the discharge delay of the MCML gate in $HIGH$ to LOW transition, while the load resistor controls the charging of the output nodes in LOW to $HIGH$ transition.

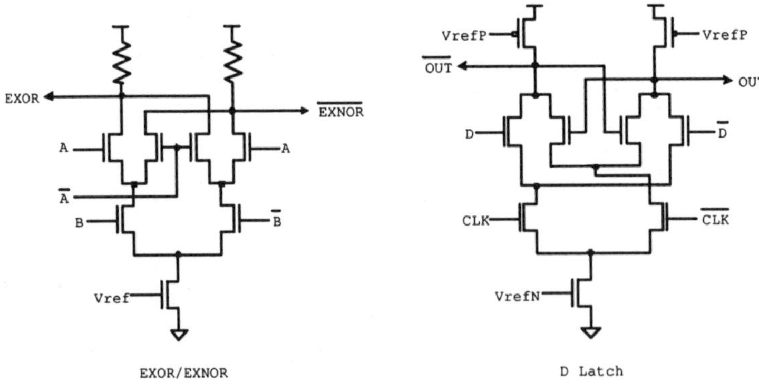

Fig. 3. MCML logic family.

MCML style can be used to construct any logic functions. Fig. 3 shows the MCML logic family. The configuration of pull-down network can be determined easily using a Binary Decision Diagram (BDD).[2] In the D latch, the pull-up transistor is implemented with a PMOS transistor controlled by a fixed gate voltage VrefP which is working in the linear region. VrefP determines the equivalent DC resistance of the PMOS load transistor.

2.2. *Performance*

Firstly, the performance of conventional CMOS circuits should be reviewed. A conventional CMOS circuit, such as a CMOS inverter, does not have large DC current, because it consists of two complementarily operating transistors. The power dissipation and the delay time of a CMOS circuit are given by:

$$P_{CMOS} = CV_{dd}^2 f\,. \tag{1}$$

$$t_{d,CMOS} = \frac{CV_{dd}}{I} = \frac{CV_{dd}}{K(V_{dd} - V_T)^{\alpha}}\,. \tag{2}$$

In above expressions, C is the load capacitance, V_{dd} is the supply voltage, f is the operation frequency, K is a constant $(= 1/2\mu C_{ox}W/L)$, V_T is the threshold voltage, α is the velocity saturation index which takes a value of 2 for long-channel devices and around 1.3 for short-channel devices, μ is mobility of electron, W is gate width, L is gate length and C_{ox} is the gate capacitance per area. The expressions above show that the power dissipation of a CMOS gate increases with its operating frequency. Decrease of supply voltage is essential to reduce the CMOS circuit power dissipation, however, it will increase delay time of CMOS gates.

Secondly, the performance of an MCML circuit should be analyzed. An MCML circuit consists of differentially operating MOS transistors and a constant current source. It can be proven that whether an MCML gate is in the steady state or transient state the sum of the current in the two branches is always a constant.[2] So, an MCML circuit consumes nearly constant power at all time. The power dissipation and the delay time of an MCML circuit are given by:

$$P_{MCML} = V_{dd} I_{cs}\,. \tag{3}$$

$$t_{d,MCML} = \frac{C\Delta V}{I_{cs}}\,. \tag{4}$$

In the above expressions, ΔV is the output voltage swing. Clearly, the power of an MCML circuit is constant with increasing operation frequency and can be reduced by lowering the supply voltage. For an MCML circuit, ΔV is generally much less than the supply voltage and of the order of several hundred millivolts, and therefore the switching speed of an MCML circuit is much faster than a CMOS one. Reducing ΔV can improve further the speed of an MCML circuit. Moreover, it is possible to reduce supply voltage of an MCML circuit without lowering operating speed, because the delay time equation doesn't include factors of supply voltage. The energy-delay product (EPD) of an MCML circuit is given by $C^2\Delta V^2 V_{dd}/I_{cs}$. It is shown that the EPD is proportional to the square of ΔV for MCML. This fact encourages further the use of very low voltage swing circuits.

As mentioned above, the switching speed of an MCML circuit is much faster than a CMOS one because of the small voltage swing and the power of an MCML circuit can be reduced by reducing the supply voltage without deteriorating the speed. For example, the maximum toggle frequency of a conventional 0.18m CMOS inverter is only about 3.5GHz, and a 0.18m MCML inverter can operate at 10GHz-frequency.[3] For 0.5m CMOS process, the delay time of an 1.5V MCML full adder with 0.5V voltage swing is only 200ps and this is three times faster than that of a 3.3V CMOS full adder and ten times faster than that of 1.5V CMOS one.[4]

The power of a CMOS circuit increases with its operating frequency, and in contrast, the power of an MCML circuit is constant with increasing operating frequency. Therefore, an MCML circuit would have low power characteristics at higher operating frequency, such as gigahertz.

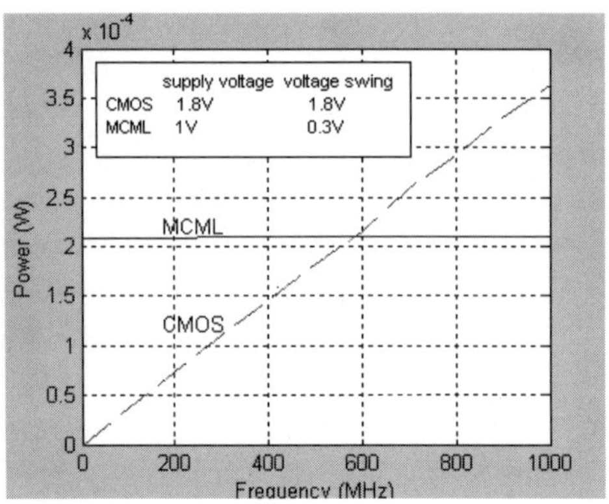

Fig. 4. Simulated power, as a function of operating frequency.

Fig. 4 shows the power dissipation of an inverter as a function of operating frequency for 0.18m process. The power dissipation of CMOS inverter at 1.8V exceeds that of a MCML inverter at 1V when operating frequency is larger than 600MHz. The power of a 1GHz MCML inverter at 1V is 57.7% of that of CMOS at 1.8V. In the deep sub-micrometer (DSM) regime, realization of high-speed circuits becomes imaginable. So, the high-speed feature of MCML circuits will become more and more remarkable in the future. In a word, MCML circuits can operate with lower signal voltage swing and higher operating frequency at lower supply voltage than conventional CMOS circuits.

By now, we have seen and analyzed an individual MCML gate. Next, a chain of cascaded gates will be explored. Let's assume that the circuit is a chain of N identical gates, that is, the logic depth of the circuit is N. The expressions for the total propagation delay, power dissipation and power-delay product of a CMOS circuit and an MCML circuit are given by:

$$t_{d,CMOS} = \frac{NCV_{dd}}{I} = \frac{NCV_{dd}}{K(V_{dd} - V_T)^{\alpha}}. \tag{5}$$

$$P_{CMOS} = NCV_{dd}^2 f = NCV_{dd}^2 f \frac{1}{t_{d,CMOS}}. \tag{6}$$

$$E_{CMOS} = t_{d,CMOS} P_{CMOS} = NCV_{dd}^2. \tag{7}$$

$$t_{d,MCML} = \frac{NC\Delta V}{I_{cs}}. \tag{8}$$

$$P_{MCML} = NV_{dd} I_{cs}. \tag{9}$$

$$E_{MCML} = t_{d,MCML} P_{MCML} = N^2 C \Delta V V_{dd}. \tag{10}$$

$$\frac{E_{MCML}}{E_{CMOS}} = \frac{N\Delta V}{V_{dd}}. \tag{11}$$

Note that the above expressions assume that the CMOS circuit is operated at a frequency equal to the reciprocal of the propagation delay.[2] The most important conclusion drawn from the above expression is that the power-delay product of an MCML gate compared with a CMOS one increases linearly with a coefficient of N. This peculiar property of MCML circuits is due to the fact that the static power dissipation of MCML circuits always exists even when the circuit is not switching. Therefore, it is very important for MCML circuits to maintain a shallow logic depth.

As we all know, dynamic logic circuits, such as Domino circuits, also have a faster switching speed than conventional CMOS circuits. However, dynamic circuits use periodic precharge and conditional evaluation operations to realize the logic functions and this fact leads to a higher switching activity than the equivalent gate implemented using other logic styles, such as MCML and CMOS. Hence, dynamic logic circuits consume more dynamic power than MCML logic circuits. Furthermore, dynamic logic also dissipates large power in order to drive the capacitance load of the clock lines, which switch at full rate. Therefore, compared with dynamic logic circuits, MCML logic circuits can save considerable power dissipation operating at the same or higher frequency.

In the high output state of a dynamic circuit, the output node is floating, so the output voltage level is very sensitive to noise and disturbances. For instance, capacitive coupling of other signals or subthreshold leakage currents might cause a charge loss at the output node which can not be recovered, and consequently the performance of a dynamic circuit will be degraded. However, a MCML circuit doesn't suffer from charge loss, because its output node is not floating. As technology move down to DSM regime, the effects of leakage current will be more pronounced (because leakage current exponentially increases as threshold voltage scales down) and the performance of dynamic logic circuits will be degraded extremely. In contrast, the impact of leakage currents could be neglected in MCML circuits, because the leakage current is contained in the bias current produced by the current source.

As mentioned above, the power of an MCML circuit can be reduced by reducing the supply voltage. The minimum supply voltage (V_{ddmin}) of an MCML inverter is the lowest voltage at which MOS transistors, which include differential pairs and a current source, can be made to operate in a saturation region. The V_{ddmin} of an MCML circuit, such as inverter, is given by $2(I/K)^{0.5} + \Delta V$, where $(I/K)^{0.5}$ is the minimum drain-source voltage of a MOS transistor to ensure it operate in a saturation region. Note that, the MCML circuits such as NAND or EXOR, which have cascaded differential pairs, need higher supply voltage than circuits which have only one differential pair such as an inverter. The V_{ddmin} of an MCML circuit can be reduced by improvements in circuit design.[4] For example, shrinking gate length L or widening gate width W can reduce $(I/K)^{0.5}$, resulting in a smaller supply voltage. Eventually, the power dissipation could be reduced. In contrast, the V_{ddmin} of an ECL circuit, which is constructed with bipolar junction transistors (BJT), can't be reduced unless using other semiconductor devices, because the V_{ddmin} of an ECL circuit is determined by the built-in potential of BJT. For example, for 0.5m CMOS and 0.8m ECL, the V_{ddmin} of an MCML inverter is 1.2V, and the V_{ddmin} of an ECL inverter is 1.8V.[4] If the driving currents are equal, the power of a 500MHz MCML circuit at 1.2V is 67% of that of an ECL circuit at 1.8V. However, the delay time of an MCML circuit at 1.2V is 1.25 times of that of an ECL circuit at 1.8V.[4] An ECL circuit is faster than the MCML circuit because of the small base capacitance and the fast switching speed of bipolar transistors. Anyway, an MCML circuit can achieve low power operation compared with an ECL circuit, as well as maintaining a relative high speed. Furthermore, an MCML circuit also has good features of MOS devices, such as low-cost fabrication and suitability for large integration.

The output signal voltage swing of an MCML circuit is much smaller than the supply voltage. A small output swing reduces the cross talk between adjacent signals. The current supplied by power supply is constant all the while, so the large precharge current and switching current appearing in a dynamic circuit and a CMOS circuit can all be eliminated. Thus, the switching noise and supply fluctuations can be greatly suppressed. Another important feature of an MCML circuit is its noise immunity due to its inherent differential nature and common mode rejection. For these reasons, MCML is suitable for mixed-signal design, because it provides an analog friendly environment and reduces the interference between the digital and analog blocks. This is very pivotal for integrating analog and digital circuits onto the same die to obtain high device density and small die area. In Ref. 1, MCML circuits are analyzed and applied to low power, low noise CORDIC computation in mixed-signal environments.

2.3. *Design consideration*

In reality, if properly designed, MCML transistors can never operate in totally cut-off region independent of the input combination. MCML transistors can only operate in the saturation and linear regions and the transistors only experience a transition from the saturation to linear region and vice versa, which takes a shorter time than from cut-off to saturation region and leads to a faster switching.[5] Note that, under this situation the high output voltage level does not approximate to V_{dd} and it experiences a voltage drop, because a little current still flows in the linear branch of the MCML gate.

Furthermore, MCML circuit optimization has more degrees of flexibility in parameter selection than many other logic styles. The only two parameters of other logic circuits (such as CMOS circuits) which can be optimized are transistor sizes and supply voltage. In contrast, MCML circuits can be optimized by adjusting the voltage swing, bias current, supply voltage, and transistor sizes. This has been a key advantage over conventional CMOS designs as process technology scales down. However, the parameters of MCML circuits tend to be tightly coupled and don't allow independent selection. So, quickly optimizing transistor-level design is a real challenge for circuit designers.

As mentioned above, a small output voltage swing of MCML reduces the cross-talk noise between adjacent signals and speeds up the MCML gate. However, an excessive small output swing is not recommended, because it caused the gate to be more susceptible to noise. Increasing pull-up resistor leads to a larger output voltage swing, but it reduces the charging current resulting in speed deterioration. Therefore, this is a speed-robustness tradeoff. Moreover, increasing current can both increase output voltage swing and speed up the MCML gate, but it increases power dissipation. This is a speed-power tradeoff. Thus, the suitable output voltage swing should be chosen to balance among power, speed and robustness.

2.4. *Characteristic and application*

MCML circuits output complementary signals and belong to differential logic style. Generally, differential logic style is area efficient in implementing arithmetic circuits and XOR based logic systems. So, as a rule of thumb, MCML style can be used to construct high-speed exclusive-or (EXOR)/exclusive-nor (EXNOR) circuits with higher area efficiency.

The power dissipation of an MCML circuit is obtained from an equation $V_{dd}I_{cs}$, and it is independent of load capacitance and remains constant as fanout increases. In contrast, the power dissipation of a CMOS circuit increases with its fanout. So, MCML circuits are suitable for application of large fanout driving circuits, such as clock drivers and bus drivers, and racing problems caused by clock skews can also be alleviated. For example, in a high-speed low power microprocessor, MCML circuits can be applied to high-speed functional blocks, such as a clock driver, registers and memories. Other parts of the processor can be implemented by CMOS circuits. The clock signals and data on bus-lines can use MCML level such as 0.5V to reduce the cross talk between adjacent signals.[4]

In the past few years, MCML has been widely used in integrated optical transceivers to realize Gbps multiplexers and demultiplexers, frequency divider and clock multiplier unit (CMU), because of its high-speed, low-power and low noise features. A 10Gb/s 8:1 multiplexer/1:8 demultiplexer in 0.18m CMOS process, a 2.5 to 10GHz CMU in 0.18m CMOS process, a 30Gb/s 1:4 demultiplexer in 0.12m CMOS process and a 25GHz static frequency divider implemented in 0.12m CMOS process using MCML have been reported.[3,6–8] The MCML multiplexer/demultiplexer ICs are not only 2 times faster than conventional CMOS one, but also consume less than 25% power than that of Si bipolar or GaAs ICs.[3] The frequency divider and logic elements of the CMU are all implemented using MCML to ensure both low sensitivity to and low generation of power supply and substrate noise.[8]

2.5. *Disadvantages*

Although MCML style tops many other logic styles in power and speed performance aspects, it still has some major drawbacks that limit its use in digital systems.

Firstly, one of the most serious drawbacks associated with MCML is the continuous static power dissipated by the constant current source. In order to eliminate the static power dissipation, some improved techniques have been devised, which will be discussed later in this paper. Secondly, MCML circuits are not suitable for power-down modes, because all of the MCML flip-flops and latches will lose their storage value if the constant current source is turned off. Thirdly, MCML is not convenient for realizing standard cells. This is partially attributed that the needs of voltage level of every input signal are potentially not the same for the gates which have cascaded differential pairs. Fourthly, as technology scales down in DSM regime, many new problems will emerge in design of MCML circuits. Saturation current of an NMOS transistor attributes to MCML's high switching characteristic. Due to short channel effects, the carrier velocity will saturate and the saturation current will decrease. Thus, the switching speed of an MCML gate will decrease and the speed enhancement of MCML will be suppressed when moving deeper in the DSM regime.

Also, as the ratio of supply voltage and threshold voltage (V_{dd}/V_T) scales down with technology scaling down, the size of the current source as well as the differential pair will increase in order to attain operation in the saturation region. This fact leads to increased area and load capacitance for previous stages, and would augment delay as well as power. Moreover, scaling down V_{dd}/V_T would limit the number of stacked levels per MCML gate.

Lastly, MCML circuits may also have some trouble with V_T variations, because the devices in a MCML gate have to be matched in order to ensure correct functionality. Deviations in V_T between differential pair transistors increase the delay time of a MCML gate, so the maximum operation frequency of a MCML gate is reduced by the V_T fluctuation.[4] Because the V_T fluctuation will increase as the gate length decreases, it becomes a serious problem for deep sub-micron MOS transistors. This is a general challenge for analog circuits operating in the DSM regime. In Ref. 3, a feedback MCML has been devised to increase the immunity to V_T fluctuation, so it is very suitable for gigahertz operation in DSM regime.

3. Improvements of Basic MCML Circuits

A serious drawback of MCML circuits is its continuous static power dissipated by the current source, therefore several newly developed circuit techniques[1,9–12] have been proposed in recent years. Those amend MCML circuits have gained significant improvement in performance. So, they become more attractive in low-power and high-speed logic design by virtue of their excellent performance.

3.1. *Dynamic current mode logic (DyCML)*

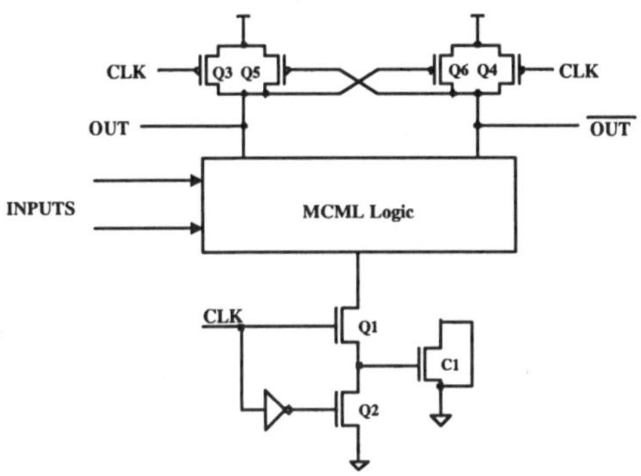

Fig. 5. Architecture of a DyCML gate.

In a DyCML gate[9] shown in Fig. 5, a dynamic current source (transistor Q1) with a virtual ground (transistor C1, which is used as a capacitor) is invented instead of the constant current source to eliminate the static power and other effects associated with the static current source. The operation of a DyCML gate is described as follows: when the clock is LOW, the precharge transistors (Q3, Q4) charge the output nodes to V_{dd}, and at the same time transistor Q2 turns ON to discharge capacitor C1 to ground level; when the clock is HIGH, transistor Q1 turns ON creating a current path from the precharged output nodes to the virtual ground (C1) and the evaluation phase starts. The virtual ground reduces the voltage swing of DyCML gates, which makes DyCML gates faster and more power efficient than traditional MCML ones.

To compare the performance of DyCML with other logic styles, six logic gates (AND, OR, XOR, MUX, AOI, FA) are implemented using five different logic families (Conventional CMOS, Complementary Pass Logic (CPL), Domino, Dynamic Differential Cascode Voltage Switch Logic (DDCVSL) and MOS Current Mode

Logic (MCML)) and simulated using 0.6um CMOS technology at 100MHz using the 3.3V supply voltage in Ref. 9. The simulation results show that DyCML gates are the fastest among all the logic styles. For complex gates (XOR, MUX, FA), DyCML gates have the smallest power dissipation because DyCML is more suitable for complex logic gates, where the power overhead of the extra circuitry vanishes with respect to the power of the logic evaluation block. The high evaluation current and the reduced voltage swing contribute to DyCML's high-speed characteristic, and at the same time the DC current in the steady state is effectively eliminated. Thus, DyCML circuits achieve high speed at low-power dissipation compared with conventional MCML circuits.

In a word, DyCML circuits inherit the current mode scheme to reduce noise and enhance performance, and at the same time utilize the dynamic logic style to cancel static power dissipation. Therefore, DyCML circuits achieve high performance at reasonable power dissipation.

3.2. *Self-timed MOS current mode logic*

Another way towards eliminating the static power dissipation is self-timed MOS current ,mode logic (ST-MCML),[12] which employs a pulse generator instead of the constant bias voltage generator to control the gate of MCML's current source. The pulses generated by the pulse generator switch on the current source for a period of time when logic is evaluated; otherwise the current source is OFF. Fig. 6 shows the schematic diagram of the pulse generator consisting of an exclusive nor gate and a delay element,[12] which is a corrected version of Anis and Elmasry's work[a].

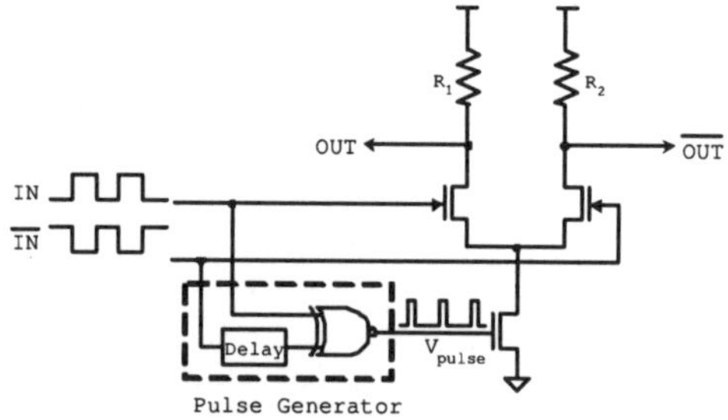

Fig. 6. ST-MCML schematic.

Use of the pulse generator eliminates the static currents when the circuit is not operating. Therefore, the power dissipation of ST-MCML circuit is proportional to

[a]The AND gate in Anis and Elmasry's work (Ref. 6) should be replaced by an EXCLUSIVE NOR gate to form a correct pulse generator, because whatever change happens with the input signal (LOW to HIGH or HIGH to LOW), the evaluation phase should be activated.

the operating frequency, and over a wide range of frequencies ST-MCML design dissipates significantly less power than conventional MCML design. Furthermore, the ST-MCML technique doesn't employ a voltage bias generator and no reference voltage distribution tree is needed, and thus the complexity of the layout stage will be decreased. However, the above merits have been obtained at the cost of good noise characteristic, because replacing the constant current source with a pulsing one would cause unwanted noise, which will interfere with other circuits. So, ST-MCML is not suitable for mixed signal designs.

The above two methods efficiently eliminate the static power dissipation, and at the same time result in a small dynamic power dissipation. On the whole, the power reduction is primary and considerable PDP reduction compared to conventional MCML circuits has been obtained.

3.3. *Adaptive current control technique*

One obstacle encountered in the production of VLSI's operating at gigahertz-order frequencies is the excessive design margin that must be reserved to accommodate circuit delay deviations. The circuit delay deviations are mainly caused by variations of device parameters (such as gate length and oxide thickness) and operating-environmental factors (such as temperature and supply voltage). In order to compensate for the circuit delay deviations and eliminate excessive static power dissipation of MCML circuits at low operating frequency, an adaptive current control (ACC) technique for MCML circuits has been proposed in Ref. 1,10,11.

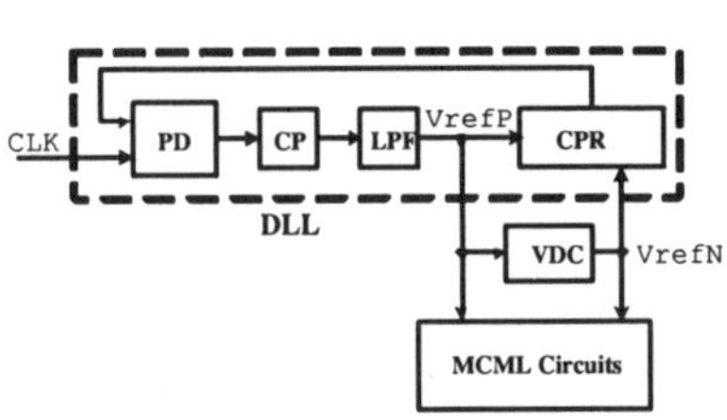

(a) Block diagram of ACC technique

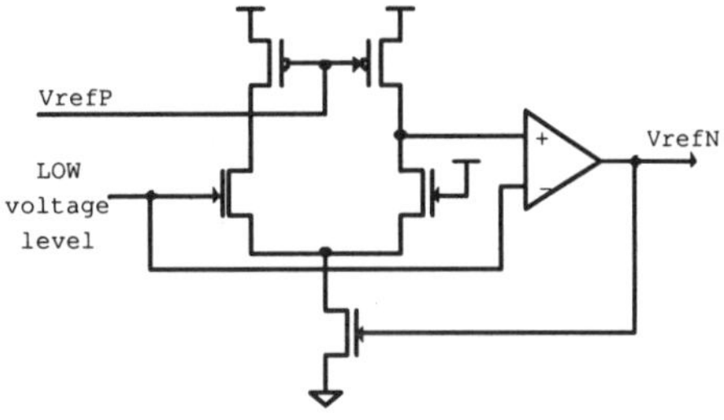

(b) Schematic diagram of Variable delay controller (VDC)

Fig. 7. Adaptive current control circuits.

The ACC technique is shown in Fig. 7,[11] where PD is a phase detector, CP is a charge pump, LPF is a low-pass filter, CPR is a critical path replica circuit, and VDC is a variable delay controller. The ACC technique constructs pull-up resistors and the constant current source with MOS transistors. By using the delay-locked loop (DLL) and feedback techniques, the ACC circuit controls the gate voltages of pull-up transistors and current source transistor automatically so as to adaptive its delay time to the variation of parameters or operating frequency while maintaining the voltage swing relatively constant for varying current. For example, if the operating frequency is reduced, the power dissipation can be reduced linearly as it is in CMOS circuits by using the ACC technique, because the drive current of all MCML gates is reduced according to the variation of operating frequency. In a word, MCML circuits using the ACC technique can save power dissipation at lower operating frequency, and guarantees the correct operation at desired frequency without influences of the device and environment variations.

If ACC technique is applied to pipelined MCML circuits, an adaptive pipeline (APL) technique using MCML circuits is obtained.[11] Because all of the gates in the pipeline must operate at a single frequency that is determined by the critical path propagation delay in just one of the pipeline stage, excessive power dissipation is caused in the relatively faster stages. By individually controlling the delay time of every MCML pipeline stage, the APL technique can eliminate the excessive power dissipation and compensate for deviations in circuit delay caused by device-parameter and operating-environment variations. Furthermore, the APL technique can also compensate for clock skew and increase design flexibility. It was shown in Ref. 11 that clock skew of 20% could be easily compensated for by using the APL technique.

In a standard CMOS design methodology, CMOS pipeline circuits must be designed to be able to operate at desired speed under the worst-case conditions. This leads to using a supply voltage which is higher than the requirement for nominal case and therefore increases power dissipation for all designs. In the real operation of CMOS circuits, a number of gates may exist under best-case conditions, and this increases unwanted noise due to wasteful speed. In contrast to CMOS circuits, MCML circuits with APL technique need only be designed for nominal case and the correct operation at all cases (worst-case, nominal-case and best-case) can be guaranteed without influences of the device and environment variations, which is attributed to the adaptive delay-adjustment function of APL technique. Furthermore, adaptive control technique can also improve the yield of circuits and allow for changes in system clock frequency after fabrication.

However, the APL technique is not perfect. Besides the extra die area spent by ACC circuit, the mistake of not picking the correct critical path will be fatal to MCML design with the APL technique. If we pick incorrect critical path during the design phase, conventional CMOS design can still operate at lower frequency. However, the MCML design with the APL technique cannot operate even at lower frequency, because a wrong critical path replica will be used as a reference during the adaptive control process.

4. Future Work

While many different facets of MCML operation have been studied, there are still many areas which require more research in the future.

The first main area of future work would be the research of automated design methodologies for MCML circuits. This automation will greatly reduce the time of implementing large logic functions. Also, design of MCML circuits should be brought into a standard digital logic design flow, which includes synthesis, placement and routing. The main challenges in this area may be the characterization of MCML circuits for the synthesis tool and the implementation of differential routing.

Secondly, with the advent of nanometer era, many new phenomena, such as short channel effects and hot carrier effect, will become obvious and greatly impact the performance and functionality of logic circuits. MCML circuits can not escape from this impact either. How to solve these upcoming problems would be a huge challenge confronted by the circuit designer in the future.

Finally, the interface circuits, which transform MCML logic level to CMOS logic level and vice versa, also need to be addressed. These interface circuits are absolutely necessary for the systems which use MCML circuits incorporated with CMOS circuits.

5. Conclusion

MCML circuits can operate at high operating frequency with lower power dissipation and lower supply voltage than many other logic styles while maintaining many other benefits, such as low cross-talk, reduction in dI/dt effects, common mode noise immunity, and process and environmental variation elimination. MCML circuits are power-efficient in high-speed situation and provide an analog friendly environment. Therefore, MCML circuits are preferred in high frequency, low power and mixed-signal applications.

While the design complexity of MCML is higher than that of conventional CMOS, significant reductions of power-delay product can be achieved in the design of high performance computation and communication circuits by using MCML. In times to come, high operating frequency will be extremely necessary and can be easily achieved by using DSM technology, so MCML's merits will become more and more remarkable in the future.

References

1. J.Musicer and J.Rabaey, "MOS Current Mode Logic for Low Power, Low Noise CORDIC Computations in Mixed-Signal Environments", in *Proc. Int. Symp. Low-Power Electronics and design*, Rapallo, Italy, July 2000, pp. 102–107.
2. Jason Musicer, "An analysis of MOS current mode logic for low power and high performance digital logic", M.Sc thesis, University of California at Berkeley, 2000.
3. Tanabe, A et.al., "0.18-um CMOS 10-Gb/s multiplexer/demultiplexer ICs using current mode logic with tolerance to threshold voltage fluctuation", *IEEE J. Solid-State Circuits* **36** (2001) 988–996.
4. M.Yamashina and H.Yamada, "An MOS Current Mode Logic (MCML) Circuit for Low-Power Sub-GHz Processors", *IEICE Trans. Electron* **E75-C** (1992) 1181–1187.
5. M.Anis, M.Allam, and M.Elmasry, "Impact of technology scaling on digital CMOS logic styles", *IEEE Transaction on Circuits and Systems* **49** (2002) 577–588.

6. H. Knapp, H.-D. Wohlmuth, M. Wurzer, M.Rest, "25GHz static frequency divider and 25Gb/s multiplexer in 0.12um CMOS", in *ISSCC'02 Digest of Technical Papers*, 2002, pp. 302–303.
7. Rylyakov, A et.al., "A 30Gb/s 1:4 demultiplexer in 0.12um CMOS", in *ISSCC'03 Digest of Technical Papers*, 2003, pp. 176–486.
8. Remco C.H. van de Beek et.al., "A 2.5 to 10GHz clock multiplier unit with 0.22ps RMS jitter in a 0.18um CMOS technology", in *ISSCC'03 Digest of Technical Papers*, 2003, pp. 178–486 .
9. Mohamed W. Allam and Mohamed I. Elmasry, "Dynamic Current Mode Logic (DyCML): A New Low-Power High-Performance Logic Style", *IEEE J. Solid-State Circuits* **36** (2001) 550–558.
10. M.Yamashina et.al, "A low-supply voltage GHz MOS integrated circuit for mobile computing systems", in *IEEE Symp. Low Power Electronics Digest of Technical Papers*, Oct. 1994, pp. 80–81.
11. M.Mizuno et.al, "A GHz MOS adaptive pipeline technique using MOS current-mode logic", *IEEE J. Solid-State Circuits* **31** (1996) 784–791.
12. Mohab H. Anis and Mohamed I. Elmasry, "Self-timed current mode logic for digital applications", in *IEEE International Symposium on Circuits and Systems*, May 2002, pp. V-113–V-116 .
13. Ayman H. Ismail and Mohamed I. Elmasry, "A low power design approach for MOS current mode logic", in *Proc. IEEE International SOC conference*, Sept. 2003, pp. 143–146.
14. Cao, J et.al., "OC-192 transmitter and receiver in standard 0.18-um CMOS", *IEEE J. Solid-State Circuits* **37** (2002) 1768–1780.
15. Jan Rabaey, *Digital Integrated Circuits: A Design Perspective*, Prentice Hall, 1996.
16. M. Alioto and G. Palumbo, "Modeling and Optimized Design of Current Mode MUX/XOR and D Flip-Flop", *IEEE Transaction on Circuits and Systems* **47** (2000) 452–461.
17. M. Alioto and G. Palumbo, "Design of CML gate with the best propagation delay", in *IEEE International Conference on Electronics, Circuits and Systems*, Sept. 1998, pp. 287–290.
18. Payam Heydari, "Design and Analysis of Low-Voltage Current-Mode Logic Buffers", in *Proc. Fourth International Symposium on Quality Electronic Design*, March 2003, pp. 293–298.
19. Payam Heydari and Ravi Mohavaelu,"Design of ultra high-speed CMOS CML buffers and latches", in *Proc. the 2003 International Symposium on Circuits and Systems*, May 2003, pp. II-208–II-211.
20. Massimo Alioto and Gaetano Palumbo,"CML and ECL: Optimized Design and Comparison", *IEEE Transaction on Circuits and Systems* **46** (1999) 1330–1341.
21. Navid Foroudi, Stefan Fulga, Puhalrajan Suppiah and John N.M.Peirce, "Low-voltage low-power topology for high-speed applications", in *Proc. the 2001 Bipolar/BiCMOS Circuits and Technology Meeting*, 30 Sept.–2 Oct. 2001, pp. 135–138.
22. Guzinski, Pawlowski, "Current-mode digital circuits operating in mixed analogue-digital systems", *Electronics Letters* **34** (1998) 1285–1286.

International Journal of High Speed Electronics and Systems
Vol. 15, No. 3 (2005) 615–641

World Scientific
www.worldscientific.com

RECENT PROGRESS IN 40- TO 100-GBIT/S-CLASS OPTICAL COMMUNICATIONS ICS USING INP-BASED HBT TECHNOLOGIES

Kiyoshi ISHII, Hideyuki NOSAKA, Kimikazu SANO, Minoru IDA, Kenji KURISHIMA, Tsugumichi SHIBATA, and Takatomo ENOKI

NTT Photonics Laboratories, NTT Corporation,
3-1, Morinosato Wakamiya, Atsugi-shi, Kanagawa, 243-0198 Japan
ishii.kiyoshi@lab.ntt.co.jp

This paper describes our InP-based heterojunction bipolar transistor (HBT) technologies and circuit design techniques for small-scale-integration (SSI) and medium-scale-integration (MSI) circuits for 40- to 100-Gbit/s-class optical communications systems. The circuits include a sub-4-ps emitter-coupled logic (ECL) gate, 100-Gbit/s selector circuit, 90-Gbit/s decision circuit, 50-Gbit/s 4:1 multiplexer (MUX) and 1:4 demultiplexer (DEMUX) ICs, over-40-Gbit/s 16:1 MUX IC, and 40-Gbit/s full-clock-rate DEMUX with a clock and data recovery (CDR) circuit. This paper demonstrates that InP-based HBT technologies and our circuit design techniques are attractive for fabricating ultrahigh-speed SSI circuits with data rates approaching 100 Gbit/s and low-power MSI circuits with data rates of over 40 Gbit/s.

Keywords: InP-based HBT; ECL gate; MUX; DEMUX; Decision circuit; Selector circuit; CDR; Optical communications.

1. Introduction

InP-based heterojunction bipolar transistors (HBTs) are particularly attractive for fabricating high-performance integrated circuits (ICs) because they offer high internal gain, high driving capability, high cutoff frequency fT, and high maximum oscillation frequency fmax. On the other hand, there are strong demands for more transmission capacity in optical communications systems to support various communication services. High-speed low-power ICs are necessary for broadband optical communications systems to integrate complex functions in the system, compactly and cost effectively. Recently, we have developed InP-based HBT technologies for achieving high-performance small-scale-integration (SSI) and medium-scale-integration (MSI) circuits. One device is an InP/InGaAs HBT with an undoped emitter and the other is an InP/InGaAs double heterojunction bipolar transistor (DHBT).

Emitter-coupled logic (ECL) and current-mode logic (CML) series gates are the basic components of high-speed ICs. However, the delay time of the gate often limits the circuit performance. Very-high-speed ECL/CML ring oscillators have been demonstrated using InP-based HBTs[1] and SiGe-based HBTs[2]. One effective approach to reduce the ECL/CML gate delay time without increasing power consumption is to use a low internal voltage swing. A 4.2-ps delay has been achieved at a low internal voltage swing of 250 mV using SiGe-based HBTs[2]. However, the delay cannot be further improved by reducing the internal voltage swing because the gate does not switch properly under an extremely low internal voltage swing. We have proposed a novel approach to further improve the ECL gate delay time. One purpose of this work was to evaluate the speed performance of our ECL gate. For this purpose, we designed and fabricated 19-stage ring oscillator ICs with the conventional and proposed ECL gates using InP/InGaAs DHBT technology.

A master-slave D-type flip-flop (MS-DFF) circuit plays key roles in integrated circuits for high-bit-rate optical communications systems. However, its operating speed often limits system performance. Recently, very-high-speed MS-DFF circuits using InP-based HBTs and InP-based high-electron-mobility transistors (HEMTs) have been reported. An asymmetrical latch flip-flop and a peaking inductor technique were used to boost the circuit operating speed[3-4]. We have already proposed a frequency divider IC that uses a novel toggle flip-flop (T-FF) using Si bipolar transistors[5]. This circuit configuration boosts the operating speed of the T-FF without increasing power consumption in the circuit. In the present work, we applied this circuit technique to the design of an MS-DFF to increase its speed. We designed and fabricated decision circuits with a conventional MS-DFF core and a novel core using both undoped-emitter InP/InGaAs HBT and InP/InGaAs DHBT technologies to evaluate the performance of the novel MS-DFF.

The selector gate also plays an instrumental role in optical communication ICs. It is used in the core circuit of the time-division multiplexers (MUXs). Very-high-speed selector-type MUX ICs have been designed and fabricated using InP-based HBTs[6], SiGe-based HBTs[7], and InP-based HEMTs[8]. In our work, we applied InP/InGaAs DHBTs to this circuitry to assess the speed performance of the new transistors.

A MUX, a demultiplexer (DEMUX), and a clock and data recovery (CDR) circuit are also key components of optical communications systems and measuring equipment. Considerable work related to the design and fabrication of over-40-Gbit/s-class MUX, DEMUX, and CDR circuits has been carried out using InP-based HBTs[9-12], SiGe-based HBTs[13-17], and InP-based HEMTs[18-19]. In this work, we designed and fabricated a low-power 4:1 MUX, 16:1 MUX, 1:4 DEMUX, and full-clock-rate DEMUX with the CDR using InP/InGaAs HBTs with undoped emitters.

In this paper, we introduce our device technologies and circuit design techniques for 40- to 100-Gbit/s-class SSI and MSI circuits for future optical communications systems. In Section 2, we briefly describe our InP/InGaAs HBT technologies. In Section 3, we discuss the circuit designs of the SSI and MSI circuits that enable high-bit-rate operation with low power consumption. Some novel circuit architectures are presented. In Section 4, we explain the measurement setup and present the measured IC performances.

2. Device Technologies

Recently, we have developed new two types of InP/InGaAs HBT. Both the HBTs used in our work were grown by metalorganic vapor phase epitaxy (MOVPE) on a 3-inch semi-insulating InP substrate. The base dopant was carbon. We used a simple non-self-aligned process to make the SSI and MSI circuits. Our HBT technology integrates InP/InGaAs HBTs, metal resistors, metal-insulator-metal (MIM) capacitors, and two-level interconnections separated by a low-permittivity interlayer of benzocyclobutene (BCB).

2.1. *Undoped-emitter InP/InGaAs HBT technology*

One device we developed is an InP/InGaAs HBT with an undoped emitter[20]. The undoped-emitter structure offers higher fT than the conventional n-doped-emitter one at low collector current density, so it is potentially attractive for fabricating high-speed ICs with low power consumption. Fig. 1 shows an SEM photograph of the undoped-emitter HBT. We adopted a hexagonal-shaped emitter geometry aligned parallel to the primary flat to suppress the base leakage current around the periphery. It has a 70-nm-thick undoped InP emitter, 50-nm-thick carbon-doped InGaAs base, and 300-nm-thick InGaAs collector. The cutoff frequency fT and maximum oscillation frequency fmax are approximately 150 and 200 GHz, respectively, at a collector current density JC of 50 kA/cm2 and a collector-to-emitter voltage VCE of 1.2 V. All transistors have an emitter width of 1.0 um. We mainly used the undoped-emitter InP/InGaAs HBT technology for fabricating low-power MSI circuits.

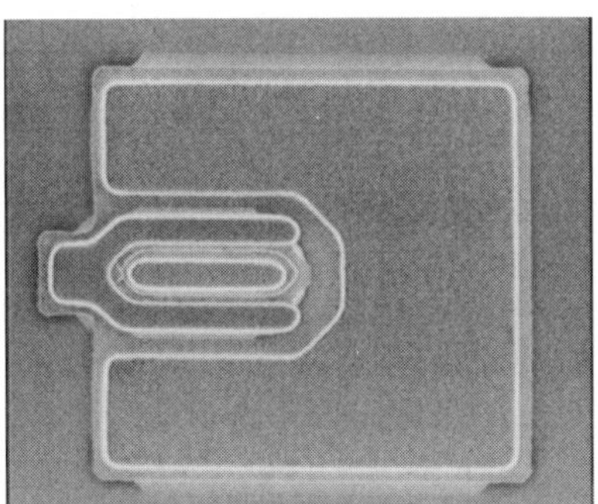

Fig. 1.

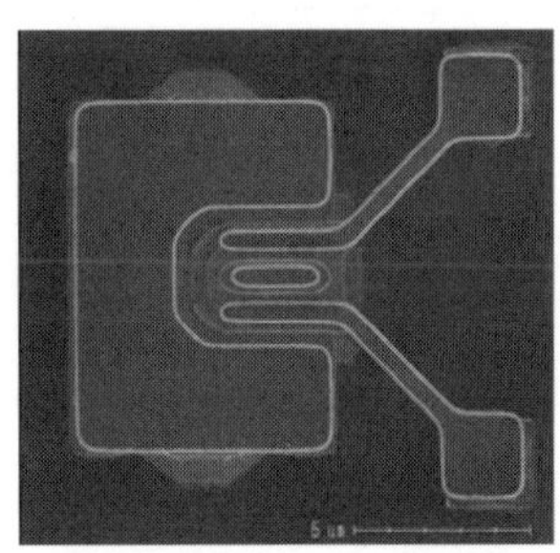

Fig. 2.

Fig. 1. SEM photograph of the undoped-emitter InP/InGaAs HBT

Fig. 2. SEM photograph of the InP/InGaAs DHBT

2.2. *InP/InGaAs double HBT technology*

The other device we developed is an InP/InGaAs double heterojunction bipolar transistor (DHBT)[21]. An SEM photograph of it is shown in Fig. 2. The detailed layer design is described in[21]. The collector thickness was designed to be 200 nm. The InP emitter layer was doped to 3 x 1017 cm-3. We introduced a base-pad isolation structure to reduce the extrinsic collector capacitance. The emitter size is 0.8 x 3 um2. The fT and fmax were measured to be 232 and 360 GHz, respectively, at JC of 2.5 mA/um2. This InP/InGaAs DHBT technology has mainly been used for fabricating ultrahigh-speed SSI circuits.

3. Circuit Design

3.1. *ECL gate design*

The circuit configuration of the proposed ECL gate is shown in Fig. 3. The feature of this circuit is the connection of the collector nodes of emitter follower transistors and load resistors. The collector nodes are connected between split load resistors R1 and R2. When R1 is set to 0 ohm, the ECL gate operates conventionally. By optimizing the values of R1 and R2, we improved the delay time. The circuit utilizes the transient current of the emitter follower to reduce the switching time of the current-switching transistors[5].

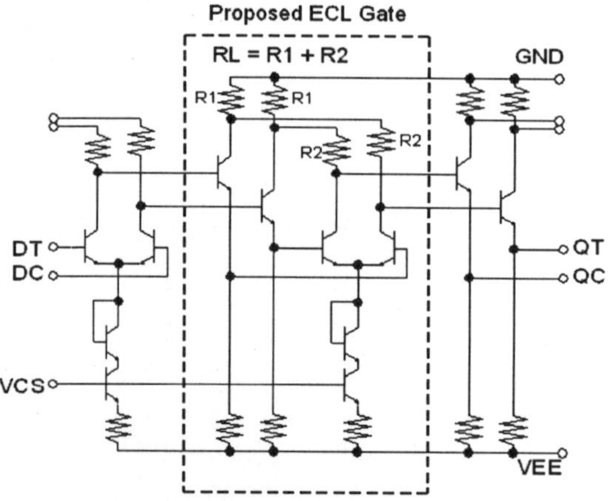

Fig. 3. Circuit configuration of the proposed ECL gate

We designed and fabricated three 19-stage ring oscillators ICs. One uses the proposed ECL gate. The internal voltage swing was designed to be 250 mV at a switching current of 6 mA (R1 + R2 = 43 ohm). The others use conventional ECL gates with an internal voltage swing of 400 mV (R1 = 0, R2 = 67 ohm) and 250 mV (R1 = 0, R2 = 43 ohm) at a switching current of 6 mA. The supply voltage VEE was designed to be –5 V. HSPICE simulation showed that the oscillation frequency of the ring oscillator using the proposed ECL gate with a 250-mV swing is about 8% higher than the conventional ECL gate with a 250-mV swing.

3.2. *Decision circuit design*

Our MS-DFF configuration is shown in Fig. 4. The important feature of this circuit is the connection of the latching circuits. Each load resistor is divided into two parts (R1 and R2) and the collector nodes of the latching circuits are connected between R1 and R2. Our sensitivity analysis using HSPICE simulation showed that the collector and base responses of the upper-level current switch transistors improve as resistor ratio R1/(R1+R2) decreases. These improvements in the responses result in a high flip-flop operating speed. The power consumption is just as low as in the conventional MS-DFF because no additional devices or power are used.

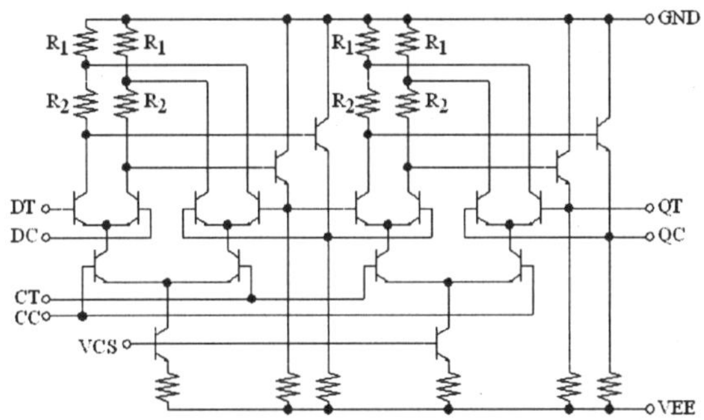

Fig. 4. Circuit configuration of the novel master-slave D-type flip-flop

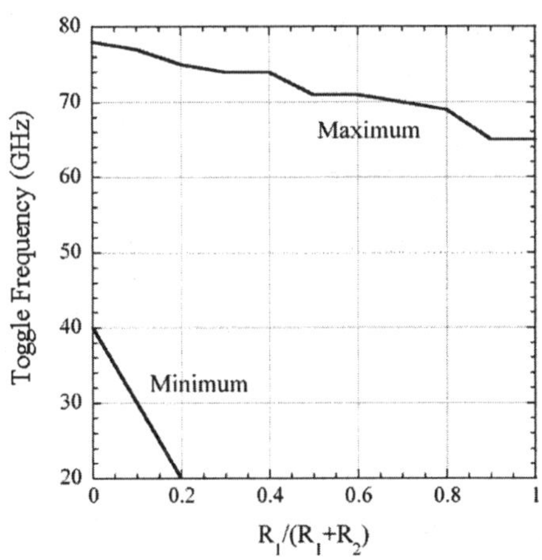

Fig. 5. Simulated maximum and minimum toggle frequencies of the novel master-slave D-type flip-flop

Fig. 5 shows the HSPICE simulation results illustrating the relationship between the maximum and minimum toggle frequencies and the resistor ratio. In this simulation, the device parameters of our undoped-emitter InP/InGaAs HBT were used[20]. For simplicity, we did not take into account the parasitic resistance and capacitance of interconnection metal wiring. The maximum toggle frequency was evaluated to be 65 GHz for the conventional MS-DFF (resistor ratio of 1.0). As the resistor ratio decreased from 1.0 to 0, the maximum toggle frequency increased up to 78 GHz.

However, at a resistor ratio below 0.2, the circuit did not operate in the low-frequency region. This is because the gain in the latching circuits was too small for the circuit to operate properly at such a low resistor ratio. As shown in Fig. 5, an approximately 14% increase in operating speed, with very-low-speed operation ensured, was achieved at a resistor ratio of 0.4.

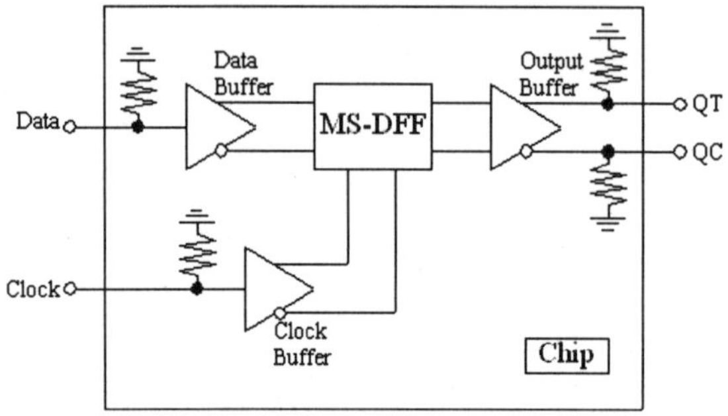

Fig. 6. Block diagram of the decision circuit

A block diagram of the decision circuit is shown in Fig. 6. The ICs consist of a data buffer and a clock buffer, the MS-DFF core, and an output buffer. A preamplifier employing the cascode connection technique was used for the data and clock buffers to obtain high-sensitivity high-speed operation. The HSPICE simulation results showed that the 3-dB-down bandwidth of the buffer was enhanced approximately 35% through the use of the cascode connection technique. We adopted on-chip terminating resistors to reduce the waveform distortion resulting from multiple reflections. A photograph of a decision circuit using undoped-emitter InP/InGaAs HBT technology is shown in Fig. 7(a). The chip size is 2 x 2 mm2. Fig. 7(b) is a photograph of a decision circuit using InP/InGaAs DHBT technology. The chip size is 1.5 x 1.5 mm2.

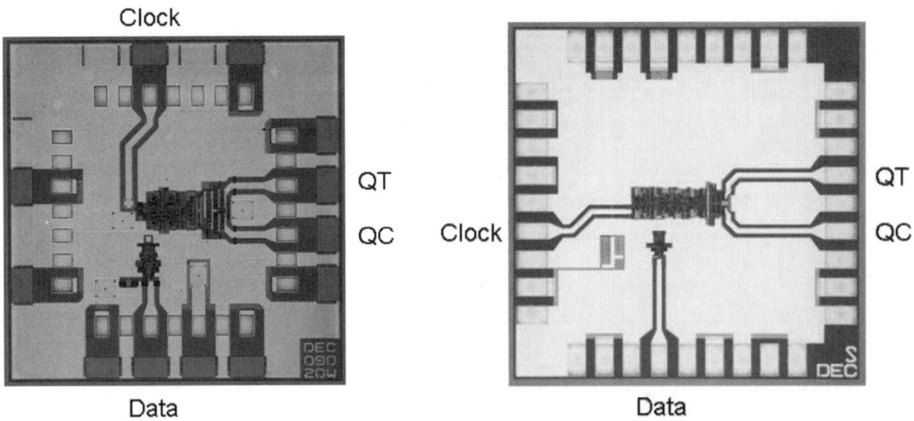

(a) (b)

Fig. 7. Chip photograph of the decision circuit. (a) Undoped-emitter InP/InGaAs HBT decision circuit. Chip size is 2 x 2 mm2. (b) InP/InGaAs DHBT decision circuit. Chip size is 1.5 x 1.5 mm2.

3.3. *Selector IC design*

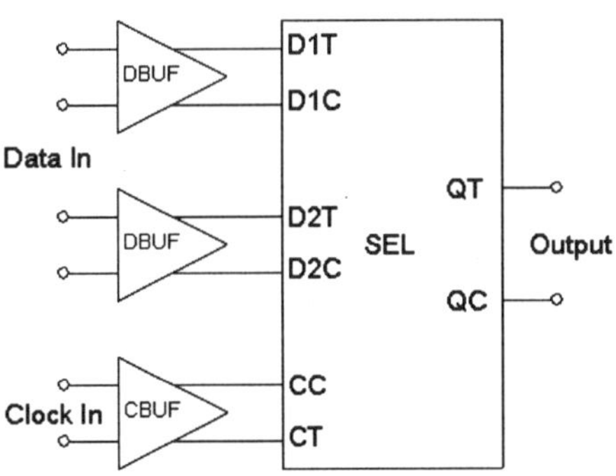

Fig. 8. Block diagram of the selector circuit

A block diagram of the selector IC is shown in Fig. 8. The IC consists of data buffers, a clock buffer, and a selector gate. The circuit configurations of the data buffer and selector gate are based on the ECL gate technique as shown in Fig. 9. The clock buffer (not shown here) is similar to the data buffer. For input impedance matching, there are 50 ohm resistors at the input of the data and clock buffers. From simulations, we found that the bandwidths of the data and clock buffers limit the whole circuit speed. Therefore, we adopted cascode connection techniques for these buffers to boost their bandwidths. The selector gate was designed to act as an output buffer that drives the load[22]. We adopted this circuit configuration because the selector gate using the HBTs has enough driving ability. This configuration lets us evaluate the speed performance of the selector gate.

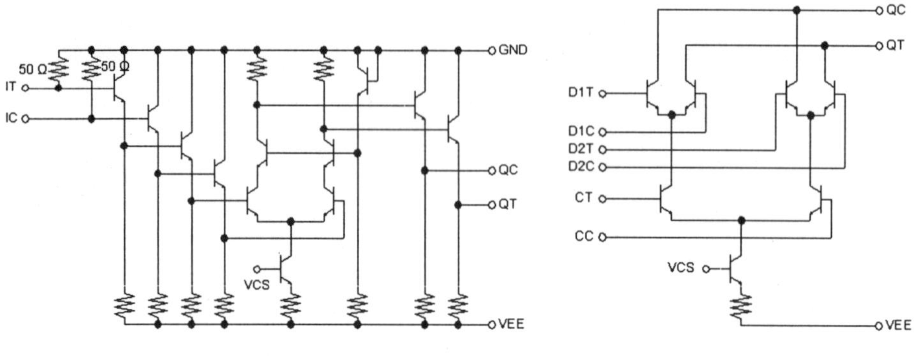

(a) (b)

Fig. 9 Circuit configuration. (a) Data buffer. (b) Selector gate

3.4. *MUX/DEMUX IC design*

3.4.1. *4:1 MUX*

A schematic of the 4:1 MUX is shown in Fig. 10. We adopt the conventional tree-type architecture. The 2:1 MUX block consists of a three-stage D-type flip-flop (TS-DFF), MS-DFF, and selector (SEL) gate to get a wide phase margin, as shown in the timing chart. The TS-DFF delays the data 2 signal by half a bit from data 1 signal. The half-bit delay ensures a wide phase margin for selecting the data signals in the SEL gate.

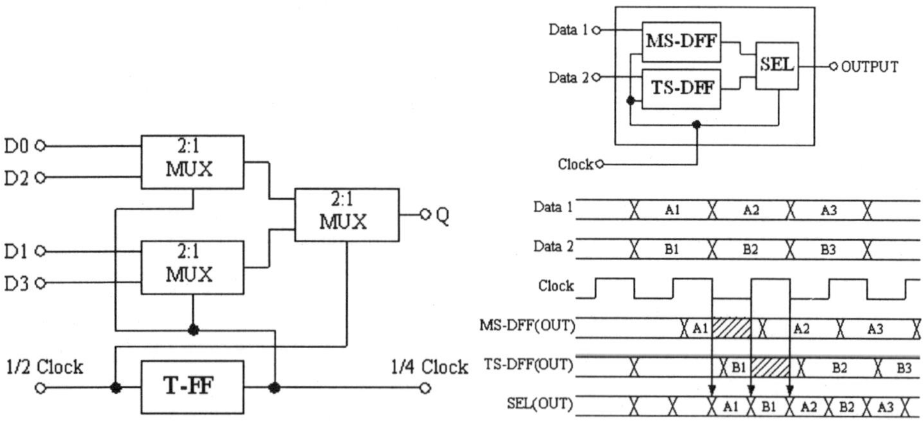

(a) (b)

Fig. 10. 4:1 MUX IC (a) 4:1 MUX architecture. (b) 2:1 MUX block and the timing chart

An ECL and a CML series gate are used for the flip-flops and SEL gate. The circuit configurations of the MS-DFF are shown in Fig. 11. In this paper, the flip-flop consisting of current switches and emitter followers is called the ECL flip-flop [Fig. 11 (a)]. The flip-flop consisting of current switches only is called the CML flip-flop [Fig. 11 (b)]. Fig. 12 shows the simulated dependence of the maximum toggle frequency on power consumption for both flip-flops, where CL is load capacitance, which includes the loading effects of subsequent devices and the capacitance of the interconnecting metal layer. We roughly assume that CL is 0.1 pF. The device parameters of our undoped-emitter InP/InGaAs HBT were used for the HSPICE simulation. The ECL flip-flop has the emitter follower between the master latch and slave one. The input impedance of the emitter follower is very high and the output impedance is very low. This results in high-speed switching of upper-level current switch transistors. In addition, the use of the emitter follower enables the upper-level current switch transistors to be appropriately biased between the base and the collector to minimize the base-collector capacitance CBC. On the other hand, the upper-level current switch transistors of the CML flip-flop are biased in the soft saturation region. Therefore, the maximum toggle frequency of the ECL flip-flop is much higher than the CML one. The ECL flip-flop is attractive for high-speed operation. Therefore, we used the ECL flip-flop in the final 2:1 MUX stage. We expect over-50-GHz flip-flop operation using our undoped-emitter InP/InGaAs HBTs.

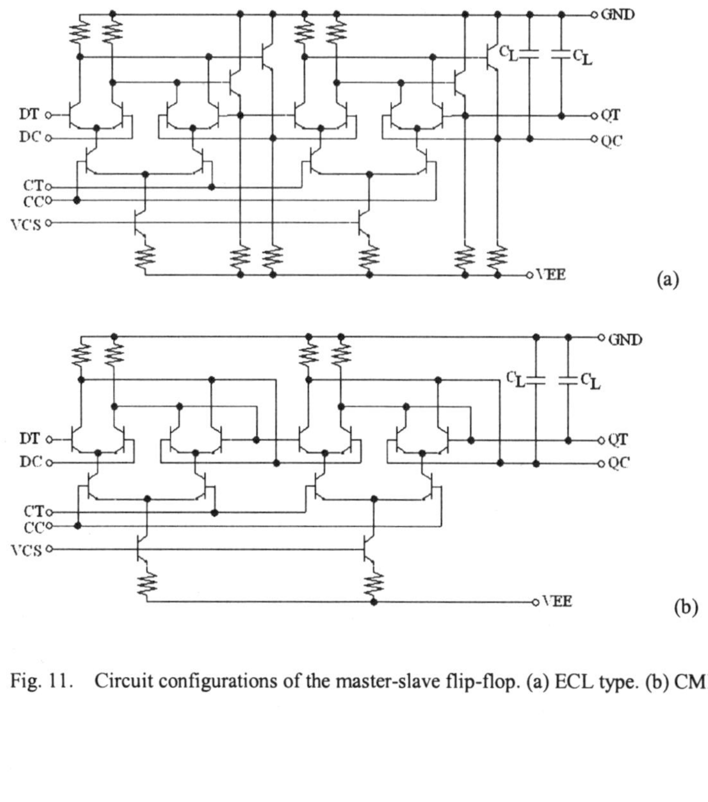

Fig. 11. Circuit configurations of the master-slave flip-flop. (a) ECL type. (b) CML type

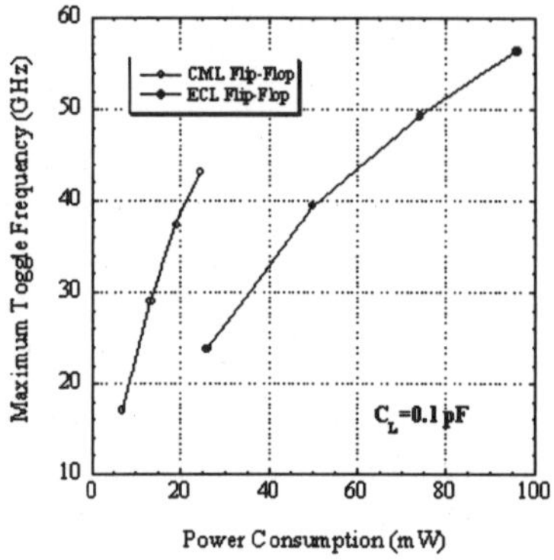

Fig. 12. Simulated dependence of maximum toggle frequencies on power consumption

On the other hand, the CML flip-flop is attractive for low-power operation at operating speeds below 40 GHz. For example, to achieve 30-GHz operation, the ECL and the CML flip-flops consume about 35 and 15 mW/flip-flop, respectively, as shown in Fig. 12. The power consumption of the CML flip-flop is less than half that of the ECL one. In order to achieve accurate operation with low power consumption, we adopted the CML flip-flop in the 4:2 MUX stages. The collector current densities (JC) of the HBTs in the CML flip-flop were designed to be about half that of the ECL flip-flop to reduce the power consumption as much as possible. The internal voltage swing was designed to be 0.5 V for both flip-flops.

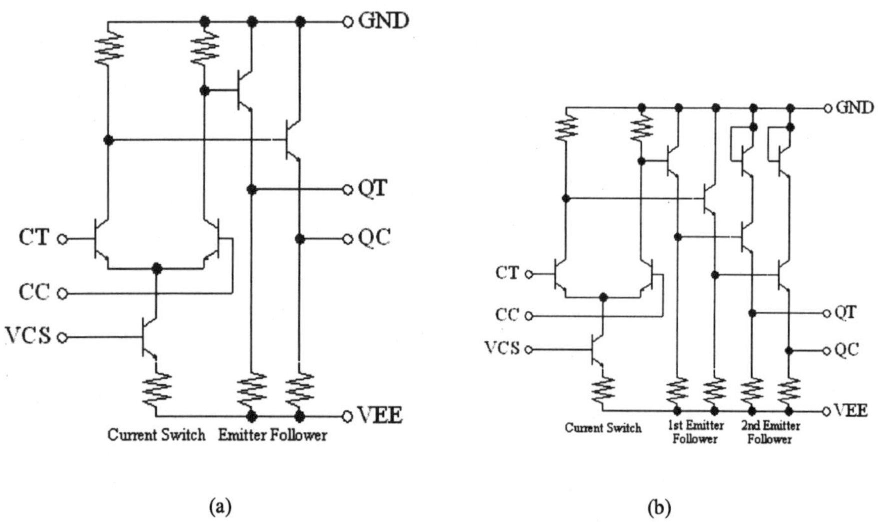

Fig. 13. Circuit configurations of the clock distribution circuits. (a) Second level output. (b) Third level output

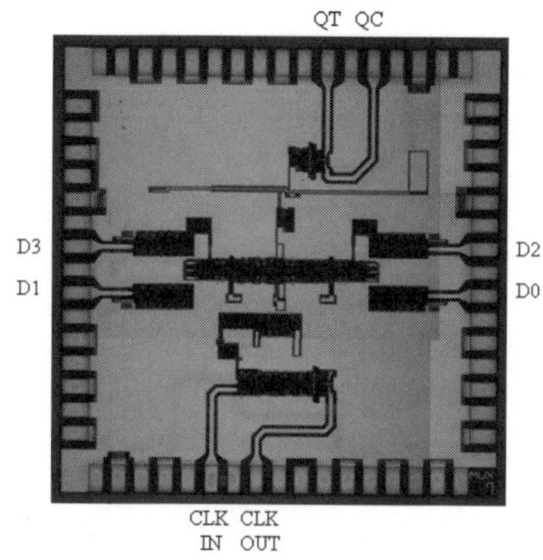

Fig. 14. Chip photograph of the 4:1 MUX

Other key blocks for achieving both high-speed operation and low-power consumption are the clock distribution circuits, which are shown in Fig. 13. The high current density of current-switching transistors results in high-speed switching. The JC values of current-switching transistors were optimized for the required operating speed. For the emitter follower transistors in Fig. 13(a) and the second emitter follower ones in Fig. 13(b), there is a trade-off between the driving capability and power consumption. Therefore, we optimized the number of transistors and JC values of the transistors of the emitter followers by determining the required operating speed and number of fan-outs. In this optimization, we also took into consideration the parasitic resistance and capacitance of interconnecting metal layers. The parasitics were extracted from the layout patterns and back-annotated in the design cycle. A chip photograph of the 4:1 MUX is shown in Fig. 14. The chip size is 3 x 3 mm2. The power supply voltage was designed to be –4.5 V.

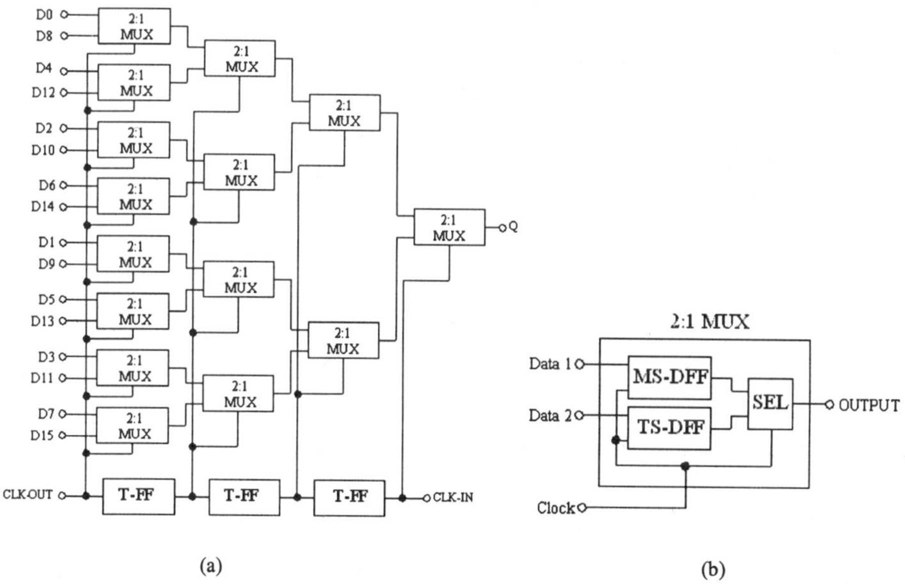

Fig. 15. (a) 16:1 MUX architecture and (b) 2:1 MUX block

3.4.2. *16:1 MUX*

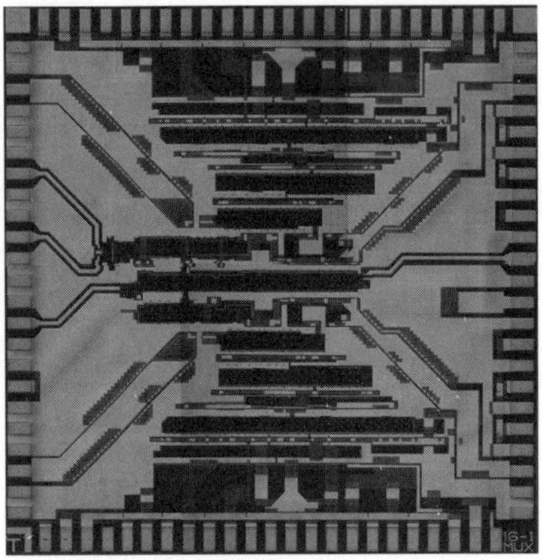

Fig. 16. Chip photograph of the 16:1 MUX.

A schematic of the 16:1 MUX IC is shown in Fig. 15. We also adopt the conventional tree-type architecture. The 16:1 MUX IC consists of fifteen 2:1 MUX blocks and three frequency dividers, which use toggle flip-flops (T-FFs). A chip photograph of the 16:1 MUX is shown in Fig. 16. The 16:1 MUX IC consists of about 1350 HBTs and 830 resistors. The chip size is 4 x 4 mm2. The power supply voltage was designed to be –4 V.

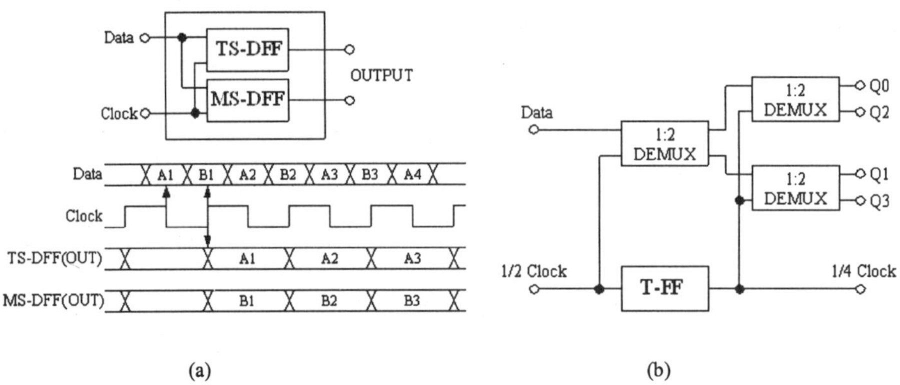

Fig. 17. 1:4 DEMUX IC. (a) 1:4 DEMUX architecture. (b) 1:2 DEMUX block and the timing chart

3.4.3. *1:4 DEMUX*

A block diagram of the 1:4 DEMUX is shown in Fig. 17. The 1:4 DEMUX employs the conventional tree-type architecture. The 1:2 DEMUX consists of a TS-DFF and MS-DFF. As shown in the timing chart, a wide phase margin is obtained by using the TS-DFF. The 1:2 DEMUX and the 2:4 DEMUX stages use flip-flops based on ECL and CML series gates, respectively. The JC of each transistor for the clock distribution circuits in the 1:4 DEMUX is also optimized. The circuit design principle is much the same as for the 4:1 MUX. A chip photograph of the 1:4 DEMUX is shown in Fig. 18. The chip size is 3 x 3 mm^2. The power supply voltage was designed to be –4.5 V.

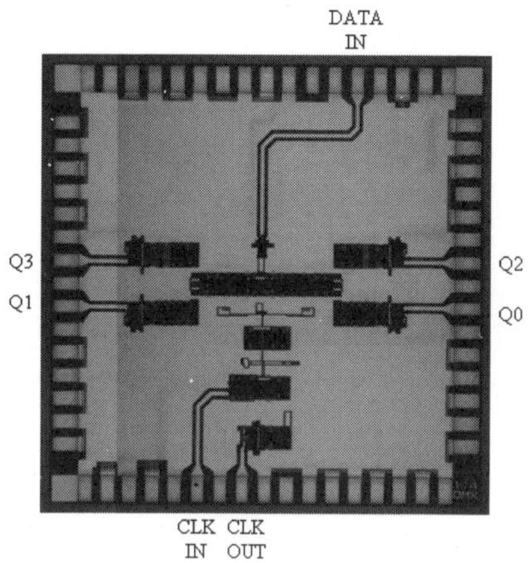

Fig. 18. Chip photograph of the 1:4 DEMUX

3.4.4. *Full-clock-rate DEMUX with the CDR design*

The one-chip full-clock-rate 1:4 DEMUX with CDR monolithically integrates a linear-type phase detector (PD), a lag-lead low-pass filter (LPF), full rate voltage-controlled oscillator (VCO), 1:4 DEMUX, and toggle flip-flop (T-FF) as shown in Fig. 19. The 1:4 DEMUX mentioned in section 3.4.3 is used as the DEMUX in this IC. The linear-type PD enables low-power operation because it can be constructed with fewer transistors than other types of PDs. It offers a wide pull-in range without a frequency acquisition circuit.

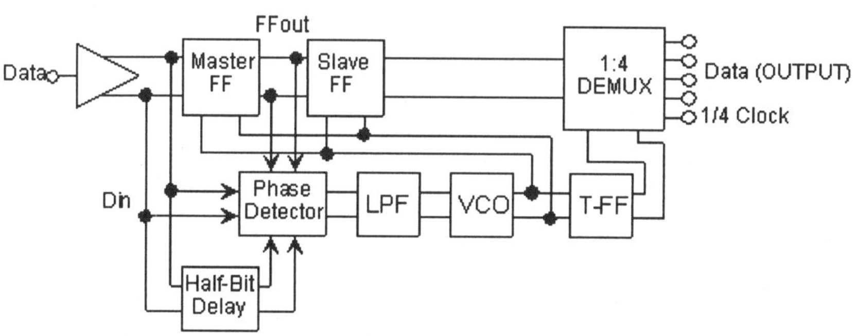

Fig. 19. One-chip full-clock-rate 1:4 DEMUX with the CDR

The half-bit-delayed data is used as standard timing data. The phase detector outputs a signal pulse that includes phase error by comparing Din, FFout, and half-bit-delayed data. The phase error signal is filtered by the LPF, which suppresses the high-frequency signal components. The LPF output signal controls the VCO. The VCO outputs a 40-GHz clock signal. This IC was designed to operate at the full clock rate of 40 GHz.

The configuration of the phase detector is shown in Fig. 20. It consists of two multiplier circuits to improve tolerance to data signal mark ratio variations[23] and data transition density variations. One multiplier detects phases between the input data Din and FFout by calculating Din x FFout. The calculated data, however, contains data-transition-density information. The other multiplier detects only the data-transition-density information by calculating Din x half-bit-delayed Din. The phase detector roughly compensates for the effect of the data transition density on the dc level of the phase detector output by finding the difference between the two multipliers' outputs.

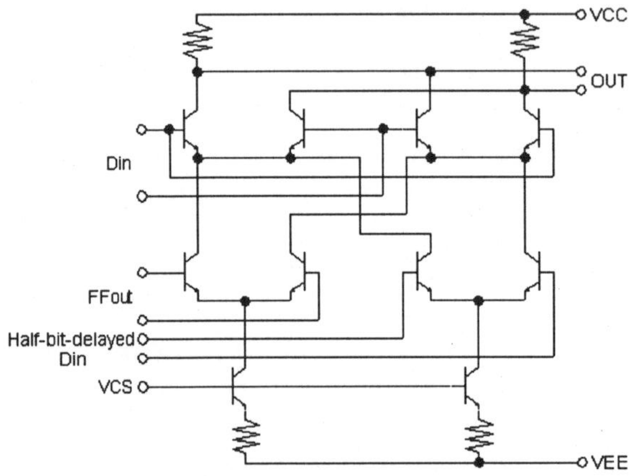

Fig. 20. Circuit configuration of the phase detector

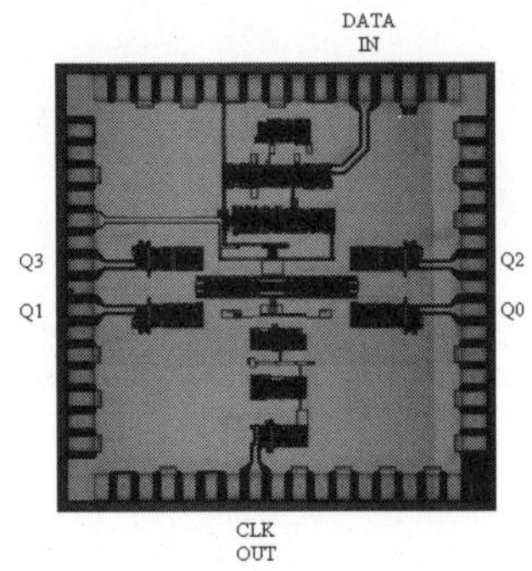

Fig. 21. Chip photograph of the 1:4 DEMUX with the CDR

A chip photograph of the full-clock-rate 1:4 DEMUX with the CDR is shown in Fig. 21. The chip size is 3 x 3 mm2. The power supply voltage was designed to be –4.5 V. The IC consists of 531 transistors and 368 resistors.

4. Measured Results

4.1. *ECL gate*

The ICs were measured on-wafer using RF probes. Fig. 22 shows the spectrum analyzer output for the 19-stage ring oscillator with the proposed ECL gate. The oscillation frequency of 8.21 GHz was obtained. The oscillation frequency corresponds to the propagation delay of 3.21 ps/gate.

Fig. 23 is propagation delay time as a function of switching current. The minimum delays of the conventional gates with the load resistor of 67 and 43 ohm were 3.67 and 3.49 ps, respectively. The 43-ohm load resistor gate was about 5% faster than the 67-ohm one. The time constants, which are related to the load resistor, are sufficiently small. Therefore, no remarkable reduction of gate delay time was anticipated below 43 ohm. In addition, as mentioned, the gate does not switch properly under extremely low load resistance.

On the other hand, the delay of the proposed ECL gate (3.21 ps) was about 8% faster than that of the conventional gate with a 43-ohm load resistor (3.49 ps). The measured results are in good agreement with the simulation ones. Both the experimental and simulation results show that the proposed ECL gate operates faster than the conventional one without increasing the power consumption.

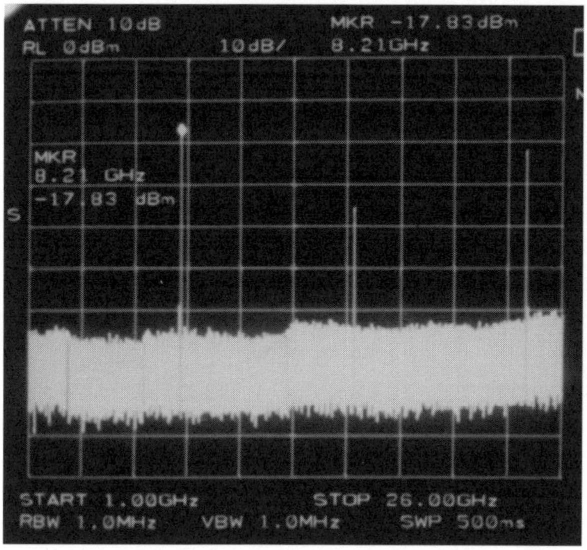

Fig. 22. Spectrum analyzer output. Oscillation frequency was 8.21 GHz

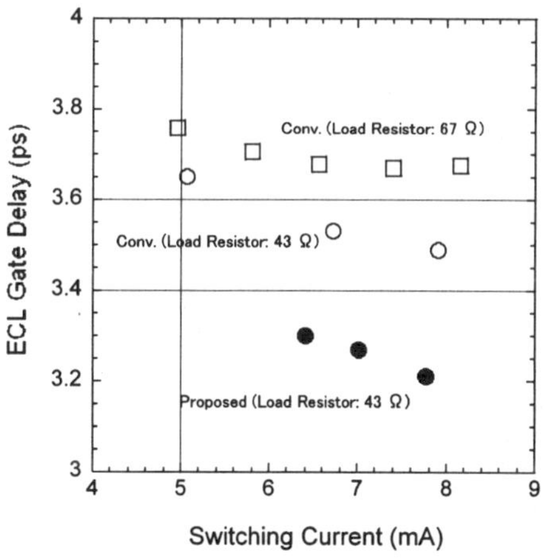

Fig. 23. Propagation delay time as a function of switching current

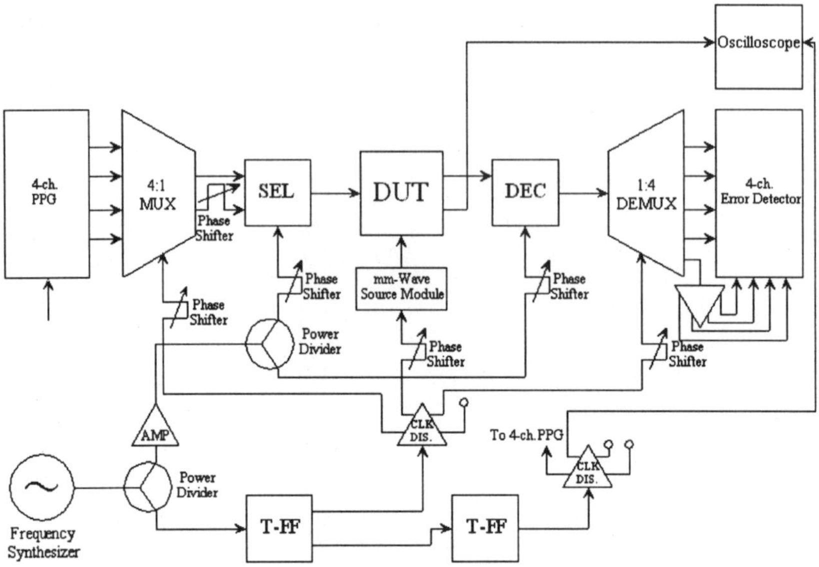

Fig. 24. Schematic diagram of the measurement system for decision circuit (>50 Gbit/s)

4.2. *Decision circuit*

The measurement setup for a bit rate of up to 50 Gbit/s is described in[24]. A schematic diagram of the measuring system used for measurements from 50 to 60 Gbit/s is shown in Fig. 24. The input data signal was generated by a 4:1 MUX module and a subsequent selector (SEL) circuit module. The output data signals of a four-channel pulse-pattern generator (PPG) were multiplexed by the 4:1 MUX and the two output data signals from the 4:1 MUX were further multiplexed by the SEL. On the other hand, one of the output data signals of the device under test (DUT) was monitored with an oscilloscope. The other one was demultiplexed into a low-speed data signal by a half-clock-operated decision circuit (DEC) module and a 1:4 DEMUX module. Error-free operation was confirmed using a four-channel error detector. The input clock signal of the DUT was generated using a mm-wave source module. The 4:1 MUX, 1:4 DEMUX, SEL, and DEC were also fabricated using our InP/InGaAs HBT technology.

For over-60-Gbit/s measurements, a true 27-1 pseudorandom bit stream (PRBS) was generated using an inhouse PPG and a subsequent InP-based HEMT selector IC module[25]. In the present measurements, the input clock was a single-ended signal.

4.2.1. *Undoped-emitter InP/InGaAs HBT decision circuits*

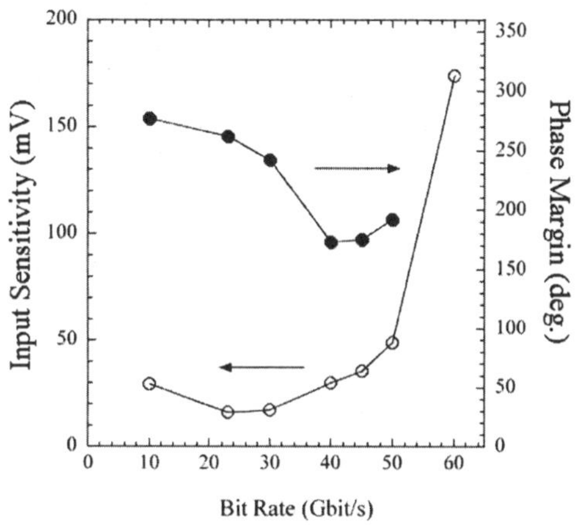

Fig. 25. Input sensitivity and clock phase margin

First, we describe the experimental results for undoped-emitter InP/InGaAs HBT decision circuits. Fig. 25 shows the dependence of the input sensitivity and clock phase margin on the data bit rate of the decision circuit with the novel MS-DFF core. We confirmed error-free operation at a data rate of up to 60 Gbit/s. Input sensitivities were approximately 50 mV at 50 Gbit/s and 170 mV at 60 Gbit/s. A wide clock phase margin of approximately 193 degrees at 50 Gbit/s was obtained. Error-free operation at a low data rate of 1 Gbit/s was also confirmed. In contrast, the decision circuit with the conventional MS-DFF core operated at a data rate of only up to 52 Gbit/s. The decision circuit using the novel MS-DFF was approximately 15% faster than the circuit using the conventional one. This result is in good agreement with the HSPICE simulation. The power consumption was approximately 0.7 W at the designed supply voltage of –4.5 V for both decision ICs.

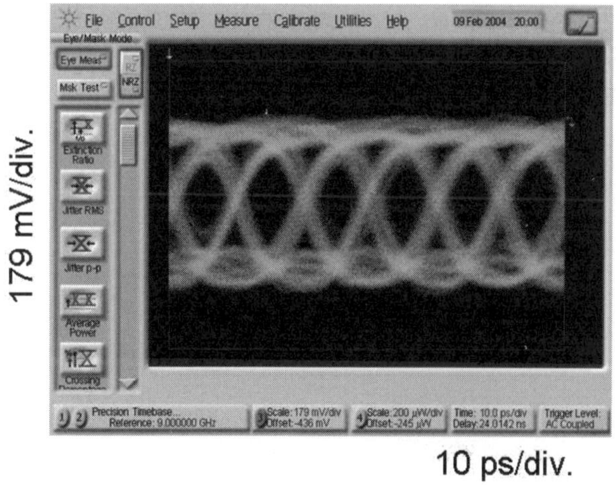

Fig. 26. Output eye pattern without a precision time base at a data rate of 60 Gbit/s

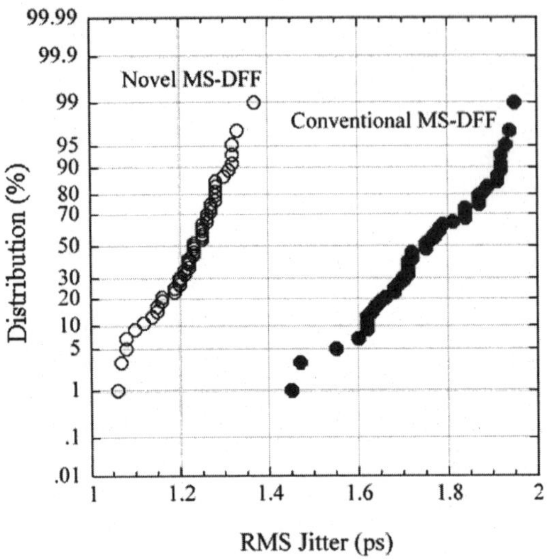

Fig. 27. Comparison of RMS jitter for output eye diagrams at a data rate of 50 Gbit/s

The output pattern at 60 Gbit/s is shown in Fig. 26. It exhibits good eye opening. A precision time base was not used for the output eye pattern observation at 60 Gbit/s because it was not available for this operating speed. The output data swing was approximately 0.85 Vp-p. The rise and fall times (20-80%) were approximately 10.3 and 7.1 ps, respectively. Fig. 27 shows the RMS jitter of the output data signals at the bit rate of 50 Gbit/s for the decision ICs with the novel and conventional MS-DFF cores. We measured 50 chips for both ICs. The average values of RMS jitter for the decision IC with the novel MS-DFF and conventional one were approximately 1.22 and 1.75 ps, respectively. It is obvious that the novel MS-DFF has better jitter performance. The output jitter was suppressed using our circuit technique at a high-bit-rate of 50 Gbit/s. This result also shows that the novel MS-DFF has good speed performance. Fig. 28 shows the typical output eye pattern with the precision time base at 50 Gbit/s for the decision circuit with the novel MS-DFF core. Our decision circuit yielded a jitter-suppressed 50-Gbit/s output data signal.

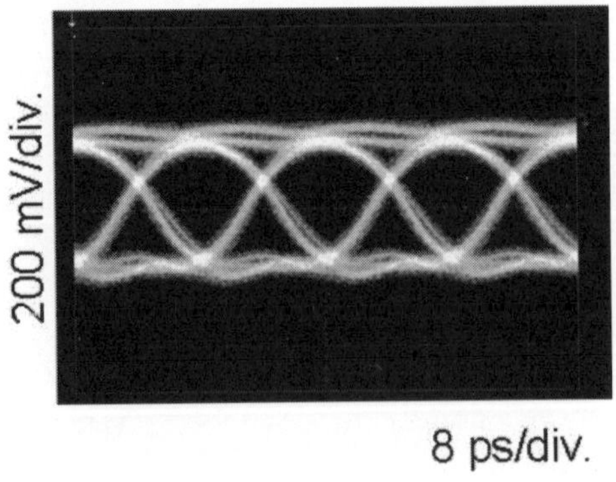

Fig. 28. Output eye pattern with a precision time base at a data rate of 50 Gbit/s

4.2.2. *InP/InGaAs DHBT decision circuits*

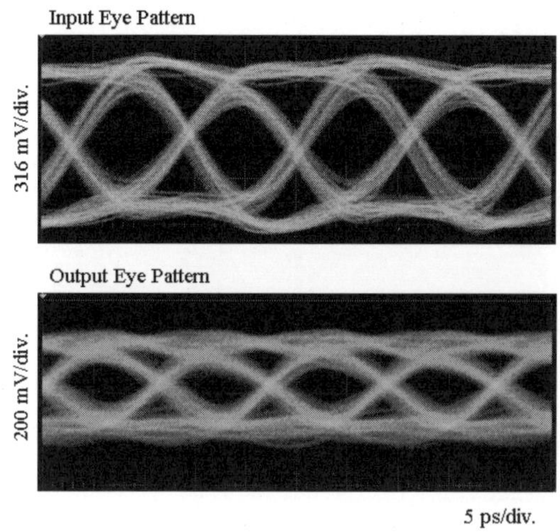

Fig. 29. Input and output eye patterns at a data rate of 90 Gbit/s. Word length is 27-1

Next, we describe the experimental results for InP/InGaAs DHBT decision circuits. The decision circuit with a conventional MS-DFF core operated at a data rate of up to 75 Gbit/s. In contrast, the decision circuit with the novel MS-DFF core operated at a data rate of up to 90 Gbit/s, approximately 20% faster than the one with the conventional MS-DFF core. The input and output eye patterns at 90 Gbit/s are shown in Fig. 29. Clear eye openings were observed up to 90 Gbit/s. The rise and fall times (20-80%) were < 6.8 ps. The RMS jitter was very low: approximately 780 fs.

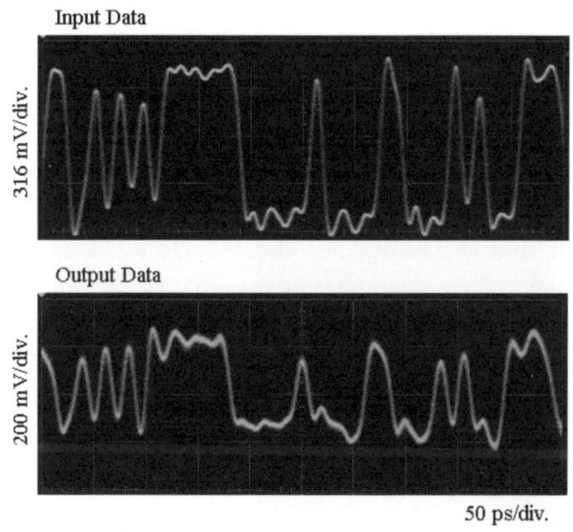

Fig. 30. Input and output data at a data rate of 90 Gbit/s. Word length is 27-1

Fig. 30 shows part of the input and output data signals with word length of 27-1 at a data rate of 90 Gbit/s. The bit sequence was correctly regenerated. The power supply voltage was –3.8 V. This low power supply voltage is allowed by the low base-emitter turn-on voltage of the InP/InGaAs HBT. The power consumption was as low as 0.5 W, including that of the clock, data, and output buffers.

4.3. *Selector IC*

The output eye pattern at 100 Gbit/s of the selector IC is shown in Fig. 31. We achieved 100-Gbit/s operation of the selector gate. The eye openings are clear, with 200-mV output swing and 400-500-fs RMS jitter.

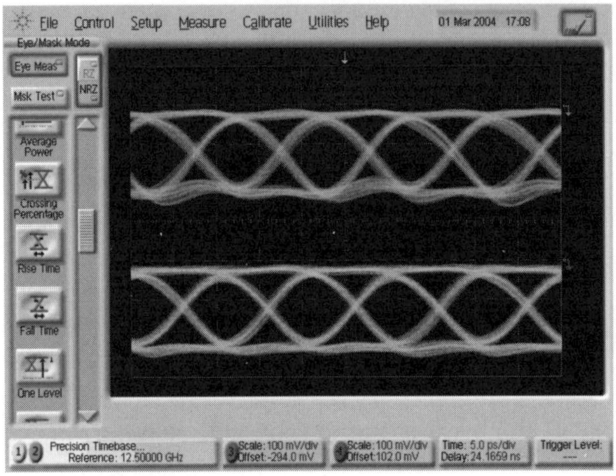

100 mV/div. 5ps/div

Fig. 31. Output eye patterns at a data rate of 100 Gbit/s (the upper is QT, the nether is QC)

4.4. *MUX/DEMUX IC*

Measurements of MUX and DEMUX ICs were also performed on-wafer using RF probes. Schematic diagrams of the measuring systems are shown in Fig. 32, where (a) is the measuring system for the 4:1 MUX and (b) is that for the 1:4 DEMUX and the full-clock-rate 1:4 DEMUX with the CDR. For measurement of the 4:1 MUX, four input data signals up to 12.5 Gbit/s with a word length of 231-1 were generated by a 4-channel PPG. The output data signal (QT) of the DUT was demultiplexed into four data channels using a demultiplexer module consisting of GaAs MESFET and InP HEMT ICs[15]. Error-free operation was confirmed for every channel using a 4-channel error detector. The output data (QC) and the output clock (CLK/4) signals were monitored with an oscilloscope. We measured the clock phase margins by shifting the four input data signals using phase shifter-A.

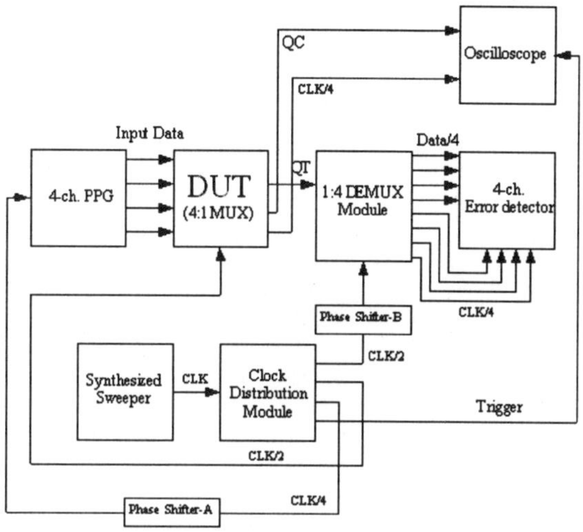

(a)

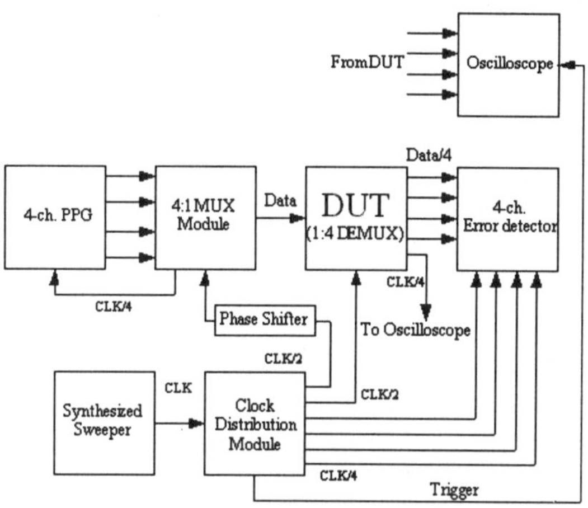

(b)

Fig. 32. Schematic diagrams of the measuring systems for (a) 4:1 MUX and (b) 1:4 DEMUX and 1:4 DEMUX with the CDR circuit

For measurement of the 1:4 DEMUX and 1:4 DEMUX with the CDR, we generated a PRBS of up to 50 Gbit/s by quadrupling a PRBS of up to 12.5 Gbit/s with a word length of 231-1 using a multiplexer module. The phase shifter was used to shift the data signal to measure the phase margin. The output data signals of the DUT were connected with the 4-channel error detector or the oscilloscope. We confirmed error-free operation for every channel in this case too. The input clock signal of the DUT (CLK/2) was not needed for the measurement of the 1:4 DEMUX with the CDR.

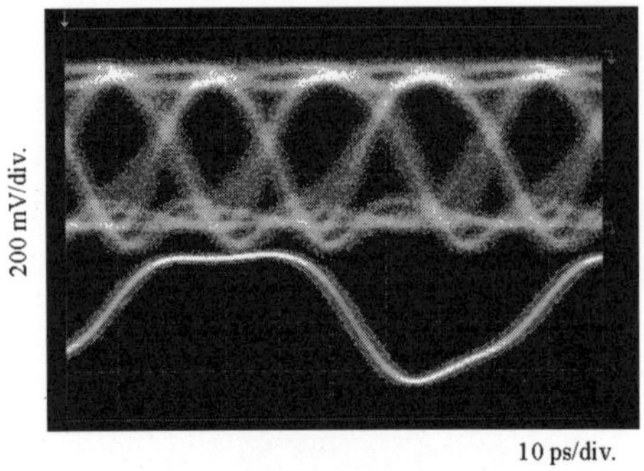

Fig. 33

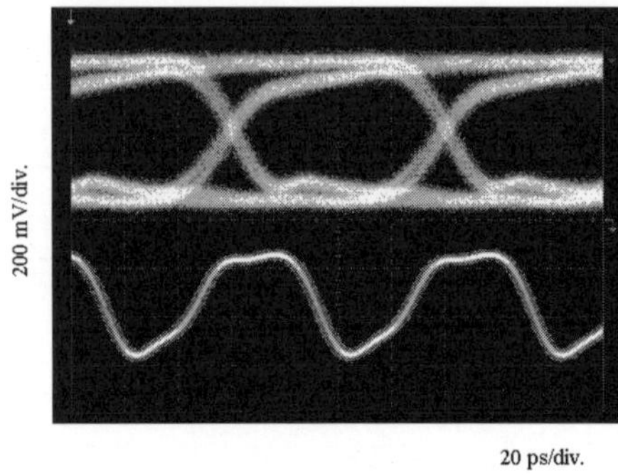

Fig. 34

Fig. 33. Output eye pattern of the 4:1 MUX at a data rate of 50 Gbit/s (upper) and output clock signal (lower)

Fig. 34. One of the output eye patterns of the 1:4 DEMUX at a data rate of 50 Gbit/s (upper) and the output clock signal (lower). Almost the same eye pattern was obtained at the other channels.

The 4:1 MUX operated with a bit error rate of less than 1 x 10-11 at a bit rate of up to 50 Gbit/s. Dead bands were not observed. Both input data and clock signal amplitudes were set to about 0.9 Vp-p. The clock phase margins were about 200 and 180 degrees at 45 and 50 Gbit/s, respectively. These phase margins include the skew for the four input data signals. The output pattern had a good eye opening at 50 Gbit/s as shown in Fig. 33. The accumulation time was 15 s. Output data swing was about 0.79 Vp-p. The rise and fall times (20-80%) were about 9.6 and 6.7 ps, respectively. The output data peak-to-peak and RMS jitters were about 5.8 and 1.1 ps, respectively.

The 1:4 DEMUX also operated with a bit error rate of less than 1 x 10-11 at a bit rate of up to 50 Gbit/s without dead bands. High input sensitivity of below 38 mV (eye height) was obtained at 50 Gbit/s. The clock signal amplitude was set to about 0.9 Vp-p. The clock phase margin was about 140 degrees at 50 Gbit/s with the data signal amplitude of about 0.9 Vp-p. A typical demultiplexed output eye pattern (channel 1) at the input data rate of 50 Gbit/s is shown in Fig. 34. Almost the same eye pattern was obtained for the other channels. The output data swing was about 0.65 Vp-p. The rise and fall times (20-80%) were about 28 and 23 ps, respectively. The output data peak-to-peak and RMS jitter were about 15 and 3 ps, respectively. The maximum operating speed of 50 Gbit/s for both 4:1 MUX and 1:4 DEMUX is limited not by the IC performances, it is limited by the measurement equipment performance. Over-50-Gbit/s operation will be achievable for both ICs.

The full-clock-rate 1:4 DEMUX with the CDR was operated at 40 Gbit/s. The 1/4 clock signal and the demultiplexed output eye patterns are shown in Fig. 35. Good eye opening was obtained. Error-free operation at 40 Gbit/s was confirmed. Fig. 36 shows the measured phase noise of the recovered 1/4 clock signal (10 GHz). The phase noise at 1-MHz off-carrier was –118 dBc/Hz.

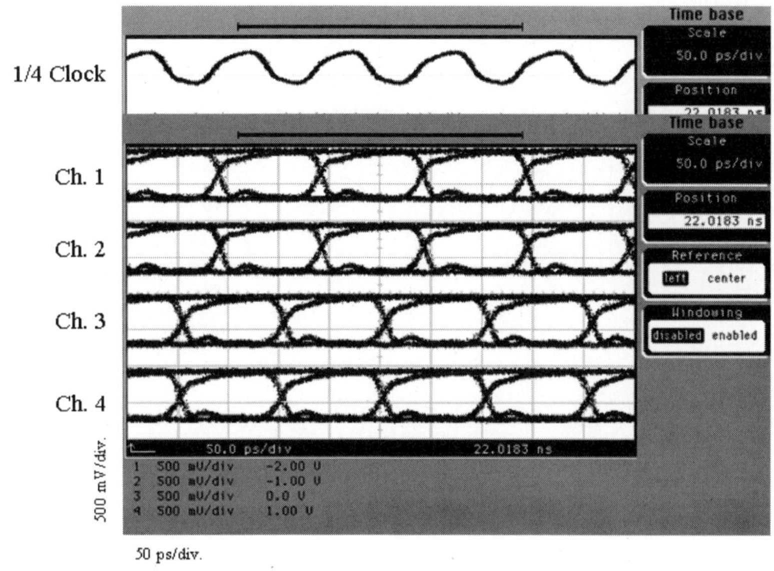

Fig. 35. 1/4 output clock signal (upper) and output eye patterns of the 1:4 DEMUX with the CDR at a data rate of 40 Gbit/s.

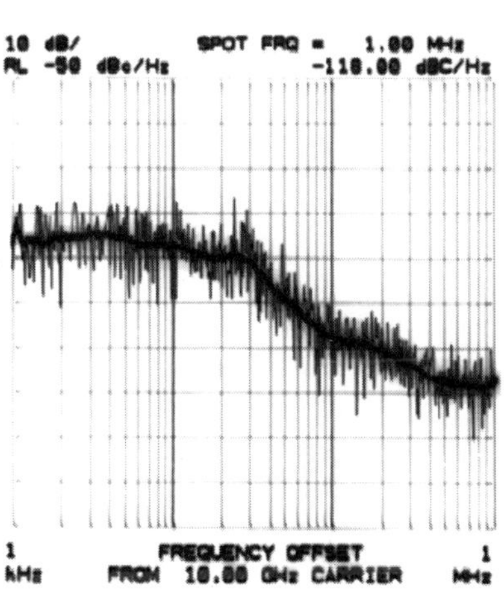

Fig. 36. Measured phase noise of recovered 1/4 clock signal for 1:4 DEMUX with the CDR

Low power consumption of about 2.3, 2.5, and 3.6 W was achieved for the 4:1 MUX, 1:4 DEMUX, and full-clock-rate 1:4 DEMUX with the CDR, respectively. Our results also indicate that a 4:1 MUX with a clock multiplier unit (CMU) circuit will be achievable by using the 4:1 MUX and the phase-locked loop (PLL) architecture of the CDR circuit in this work.

Measurements of 16:1 MUX ICs were performed on-wafer using the Cascade Microtech Pyramid Probe. A schematic diagram of the measuring system is shown in Fig. 37. The sixteen input data signals were generated from the four data signals of a PPG followed by 4:16 DEMUX module. The word length of the four-channel data was 215-1. The output data signals of the DUT were demultiplexed again into four data signals using a demultiplexer module. The bit error rates were checked for every channel with a 4-channel error detector.

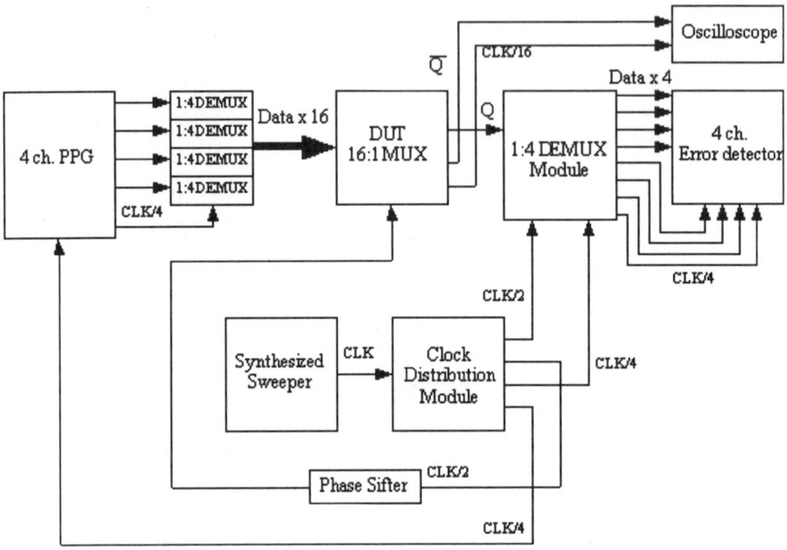

Fig. 37. Schematic diagram of the measurement system for 16:1 MUX

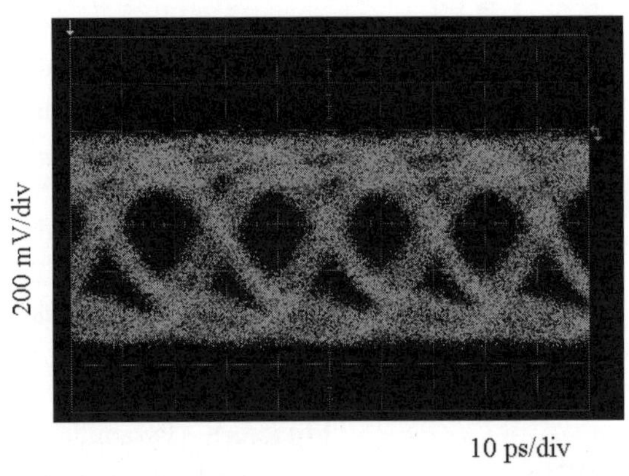

Fig. 38. Output eye pattern at a data rate of 47 Gbit/s

The 16:1 MUX IC operated at up to 47 Gbit/s. Dead bands were not observed. The output eye pattern at the bit-rate of 47 Gbit/s is shown in Fig. 38. Error-free operation was confirmed for all channels. The output pattern showed good eye opening at 47 Gbit/s. The output data swing was about 0.9 Vp-p. The rise and fall times (20–80%) were about 17 ps. The output data peak-to-peak and RMS jitters were about 8.4 and 1.5 ps, respectively. It should be noted that this performance may be limited not only by the

DUT characteristics but also by the bandwidth of the measuring equipment. The clock phase margins were about 90, 60, and 20 ps at bit rates of 40, 43, and 47 Gbit/s, respectively. These phase margins include the skew for the 16 input data signals.

The power consumption was about 3.2 W at the designed supply voltage of –4 V. When the power supply voltage was reduced to –3.65 V, error-free operation was still achieved at bit rates of up to 43 Gbit/s with low power consumption of about 2.5 W.

5. Conclusions

In summary, we have successfully designed and fabricated high-performance SSI and MSI circuits using our InP-based HBT technologies. The undoped-emitter InP/InGaAs HBT exhibits fT and fmax of approximately 150 and 200 GHz, respectively, at a low JC of 50 kA/cm2. The InP/InGaAs DHBT achieved very high fT and fmax of 232 and 360 GHz, respectively.

We introduced our circuit design techniques and some novel circuit architectures. These enable us to make a 3.21-ps ECL gate, 100-Gbit/s selector circuit, 90-Gbit/s decision circuit, low-power 50-Gbit/s 4:1 multiplexer (MUX) and 1:4 demultiplexer (DEMUX) ICs, 47-Gbit/s 16:1 MUX IC, and 40-Gbit/s full-clock-rate DEMUX with a clock and data recovery (CDR) circuit. This work demonstrates that InP-based HBT technologies and our circuit design techniques are attractive for fabricating ultrahigh-speed SSI circuits with data rates approaching 100 Gbit/s and low-power MSI circuits with data rates over 40 Gbit/s.

Acknowledgments

The authors thank M. Muraguchi, H. Sugahara and M. Tokumitsu of NTT Photonics Laboratories for their encouragement and valuable suggestions. They also thank K. Murase, K. Fujimoto, Y. Takeo, H. Taira, S. Moriyama, N. Tsuji, and S. Kurita of NTT Electronics Corporation for their support.

References

1. M. K. Srivastava, R. Raghavan, R. Thiagarajah, M. G. Case, E. Arnold, C. W. Pobanz, S. O. Nielsen, J. C. Yen, and R. A. Johnson: "InP HBT Ring Oscillator with 2.0ps/stage Gate Delay," in GaAs IC Symp. Tech. Dig., 2002, pp.171-174.
2. B. Jagannathan, M. Meghelli, A. V. Rylyakov, R. A. Groves, A. K. Chinthakindi, C. M. Schnabel, D. A. Ahlgren, G. G. Freeman, K. J. Stein, and S. Subbanna: "A 4.2-ps ECL Ring-Oscillator in a 285-GHz fMAX SiGe Technology," IEEE Electron Device Lett., 23 (2002) pp. 541-543.
3. Y. Suzuki, Y. Amamiya, Z. Yamazaki, S. Wada, H. Uchida, C. Kurioka, S. Tanaka, and H. Hida, "110Gb/s Multiplexing and Demultiplexing ICs," in Int. Solid-State Circuits Conf. Tech. Dig., 2004, pp. 232-233.
4. T. Suzuki, T. Takahashi, T. Hirose, and M. Takigawa, "A 80-Gbit/s D-type Flip-Flop Circuit Using InP HEMT Technology," in GaAs Integrated Circuits Symp. Tech. Dig., 2003, pp. 165-168.

5. K. Ishii, H. Ichino, M. Togashi, Y. Kobayashi, and C. Yamaguchi, "Very-High-Speed Si Bipolar Static Frequency Dividers with New T-Type Flip-Flops," IEEE J. Solid-State Circuits, 30 (1995) pp. 19-24
6. P. Andre, S. Blayac, P. Berdaguer, J.-L. Benchimol, J. Godin, N. Kauffmann, A. Konczykowska, A.-E. Kasbari, and M. Riet: "InGaAs/InP DHBT Technology and Design Methodology for Over 40 Gb/s Optical Communication Circuits," IEEE J. Solid-State Circuits, 36 (2001) pp. 1321-1327.
7. M. Moller, H.-M. Rein, A. Felder, and T. F. Meister: "60 Gbit/s time-division multiplexer in SiGe-bipolar technology with special regard to mounting and measuring technique," Electron. Lett., 33 (1997) pp. 679-680.
8. T. Otsuji, K. Murata, T. Enoki, and Y. Umeda: "An 80-Gbit/s Multiplexer IC Using InAlAs/InGaAs/InP HEMT's," IEEE J. Solid-State Circuits, 33 (1998) pp. 1321-1327.
9. J. P. Mattia, R. Pullela, G. Georgieu, Y. Baeyens, H. S. Tsai, Y. K. Chen, C. Dorschky, T. Winkler Von Mohrenfels, M. Reinhold, C. Groepper, M. Sokolich, L. Nguyen, and W. Stanchina, "High-Speed Multiplexer: A 50 Gb/s 4:1 MUX in InP HBT Technology," in GaAs IC Symp. Tech. Dig., 1999, pp.189-192.
10. J. P. Mattia, R. Pullela, Y. Baeyens, Y.-K. Chan, H.-S. Tsai, G. Georgiou, T. Winkler Von Mohrenfels, M. Reinhold, C. Groepper, C. Dorschky, and C. Schulien, "A 1:4 Demultiplexer for 40Gb/s Fiber-Optic Applications," in IEEE ISSCC Dig. Tech. Papers, 2000, pp. 64-65.
11. H. Nosaka, E. Sano, K. Ishii, M. Ida, K. Kurishima, T. Enoki, and T. Shibata, "A Fully Integrated 40-Gbit/s Clock and Data Recovery Circuit Using InP/InGaAs HBTs," in IEEE MTT-S Int. Microwave Symp. Dig., 2002, pp. 83-86.
12. J. Yen, M. G. Case, S. Nielsen, J. E. Rogers, N. K. Srivastava, R. Thiagarajah, "A Fully Integrated 43.2 Gb/s Clock and Data Recovery and 1:4 DEMUX IC in InP HBT Technology," in ISSCC Dig. Tech. Papers, 2003, pp. 240-241.
13. T. Masuda, K. Ohhata, N. Shiramizu, E. Ohue, K. Oda, R. Hayami, H. Shimamoto, M. Kondo, T. Harada, and K. Washio, "40 Gb/s 4:1 Multiplexer and 1:4 Demultiplexer IC Module using SiGe HBTs," in IEEE MTT-S Int. Microwave Symp. Dig., 2001, pp. 1697-1700.
14. M. Reinhold, C. Dorschky, R. Pullela, E. Ross, P. Mayer, P. Paschke, Y. Baeyens, J. P. Mattia, and F. Kunz, "A Fully-Integrated 40Gb/s Clock and Data Recovery / 1:4 DEMUX IC in SiGe Technology," in ISSCC Dig. Tech. Papers, 2001, pp. 84-85.
15. M. Meghelli, A. V. Rylyakov, and L. Shan, " 50Gb/s SiGe BiCMOS 4:1 Multiplexer and 1:4 Demultiplexer for Serial Communication Systems," in ISSCC Dig. Tech. Papers, 2002, pp. 260-261.
16. A. Koyama, T. Harada, H. Yamashita, R. Takeyari, N. Shiramizu, K. Ishikawa, M. Ito, S. Suzuki, T. Yamashita, S. Yabuki, H. Ando, T. Aida, K. Watanabe, K. Ohhata, S. Takeuchi, H. Chiba, A. Ito, H. Yoshioka, A. Kubota, T. Takahashi, H. Nii, "43Gb/s Full-Rate-Clock 16:1 Multiplexer and 1:16 Demultiplexer with SFI-5 Interface in SiGe BiCMOS Technology," in ISSCC Dig. Tech. Papers, 2003, pp. 232-233.
17. D. K. Shaeffer, H. Tao, Q. Lee, A. Ong, V. Condito, S. Benyamin, W. Wong, X. Si, S. Kudszus, M. Tarsia, "A 40/43Gb/s SONET OC-768 SiGe 4:1 MUX/CMU," in ISSCC Dig. Tech. Papers, 2003, pp. 236-237.
18. K. Sano, K. Murata, S. Sugitani, H. Sugahara, and T. Enoki, "50-Gbit/s 4-bit Multiplexer/Demultiplexer Chip-set using InP HEMTs," in GaAs IC Symp. Tech. Dig., 2002, pp. 207-210.
19. H. Kano, T. Suzuki, S. Yamamura, Y. Nakasha, K. Sawada, T. Takahashi, K. Makiyama, T. Hirise, and Y. Watanabe, "A 50-Gbit/s 1:4 Demultiplexer IC in InP-based HEMT Technology," in IEEE MTT-S Int. Microwave Symp. Dig., 2002, pp. 75-78.

20. M. Ida, K. Kurishima, H. Nakajima, N. Watanabe, and S. Yamahata, "Undoped-Emitter InP/InGaAs HBTs for High-Speed and Low-Power Applications," in Int. Electron Device Meeting Tech. Dig., 2000, pp. 854-856.
21. K. Kurishima, M. Ida, and N. Watanabe: "InP Double-Heterojunction Bipolar Transistors with a Carbon-Doped Graded Base," Extended Abstracts of the 2002 International Conference on Solid State Devices and Materials, 2002, pp. 272-273.
22. J. Hauenschild, H-.M. Rein, W. Mcfarland, J. Doernberg, and D. Pettengill: "Demonstration of retiming capability of silicon bipolar time-division multiplexer operating to 24 Gbit/s," Electron. Lett., 27 (1991) pp. 978-979.
23. N. Ishihara, S. Fujita, M. Togashi, S. Hino, Y. Arai, N. Tanaka, Y. Kobayashi, and Y.Akazawa, "3.5-Gb/s x 4-ch Si bipolar LSI's for Optical Interconnections," IEEE Journal of Solid-State Circuits, 30 (1995) pp.1493-1501
24. K. Ishii, H. Nosaka, M. Ida, K. Kurishima, T. Enoki, T. Shibata, and E. Sano, "High-input-sensitivity, low-power 43 Gbit/s decision circuit using InP/InGaAs DHBTs," Electron. Lett., 38 (2002) pp. 557-558
25. K. Murata, K. Sano, T. Enoki, H. Sugahara, and M. Tokumitsu, "InP-Based IC Technologies for 100-Gbit/s and Beyond," in Tech. Dig. of 16th International Conference on Indium Phosphide and Related Materials (IPRM'04), 2004, TuA-1-1

International Journal of High Speed Electronics and Systems
Vol. 15, No. 3 (2005) 643–665

World Scientific
www.worldscientific.com

40 Gb/s TDM SYSTEM TRANSCEIVER PROTOTYPE IN InP HBT TECHNOLOGY

KARTHIKEYAN KRISHNAMURTHY¶, JIAN XU*, XIAOQIANG SHOU, RAMAKRISHNA VETURY†, SHRINIVASAN JAGANATHAN‡, YET-ZEN LIU, RAJA PULLELA§, RUAI YU

Gtran Inc., 2651 Lavery Court,
Newbury Park, CA 91320, USA.
¶*karthik@gtran.com*

MARK J. W. RODWELL△

Department of Electrical and Computer Engineering,
University of California, Santa Barbara, CA 93106, USA.
△*rodwell@ece.ucsb.edu*

A time-division-multiplexing (TDM) system transceiver with 4–channel 10 Gb/s interfaces to achieve 40 Gb/s non-return to zero (NRZ) transmission is presented. The front-end components are implemented in an indium phosphide double hetero-junction bipolar transistor (InP DHBT) technology. They include a 4:1 multiplexer with a voltage controlled oscillator and clock multiplication unit, modulator drivers for electro-absorption and differential lithium-niobate modulators, trans-impedance amplifier, limiting amplifier and 1:4 demultiplexer with clock and data recovery circuit. The transceiver was used to test optical transmission over a 2.26 km link with a 40 Gb/s, 2^{31}–1 pseudo-random bit sequence (PRBS) data. The transmitter achieves better than 12.9 dB extinction ratio, 1 ps added root mean square (RMS) jitter and 4 dBm output power. Without any optical amplification the receiver achieves –9.1 dBm back to back sensitivity at a bit error rate (BER) of 10^{-12} and a high dynamic range of 12.5 dBm.

Keywords: Optical communication; Time division multiplexing; Multiplexing; Demultiplexing; Optical receivers; MMICs.

1. Introduction

Commercial deployment of 40 Gb/s system depends on the availability of low cost integrated circuit (IC) solutions that enable transponder manufacturers to meet cost, performance, power and size constraints. Key front-end ICs have been demonstrated in the past using HBT (SiGe, InP, GaAs) and HEMT (InP, GaAs)

*J. Xu is now with Skyworks Inc., Newbury Park, CA 91320, USA.
†R. Vetury is now with RF Micro Devices, Greensboro, NC 27409, USA.
‡S. Jaganathan is now with Texas Instruments, Dallas, TX 75243, USA.
§R. Pullela is now with Skyworks Inc., Irvine, CA 92612, USA.

technologies[1,2]. Among them InP DHBT (double h ero-junction bipolar transistor) technology exhibits excellent high-speed performance, high breakdown voltage, large-scale integration, good uniformity and reliability. This makes them suitable for realizing both broadband analog amplifiers and large-scale digital circuits required for the front-end components in 40 Gb/s systems.

In this paper we present a 40 Gb/s time-division-multiplexing (TDM) system transceiver with a 4–channel 10 Gb/s interface. The transceiver employs InP DHBT based front-end IC's, a continuous wave (CW) laser modulated by a lithium-niobate modulator, and micro-controller driven control circuitry. The transmit chip-set includes a 4:1 multiplexer (MUX) with an on-chip clock multiplication unit (CMU) and modulator drivers suitable for driving electro-absorption (EA) and differential electro-optic (EO) modulators. The receive chip-set includes a p–i–n photo-diode (PIN), trans-impedance amplifier (TIA) and a limiting amplifier (LA) integrated in a single package and a 1:4 demultiplexer (DEMUX) with an integrated clock and data recovery (CDR) circuit. Section 2 outlines the TDM system configuration. Details of the chip-set are presented in section 3. Transceiver performance is discussed in section 4. System tests for a 2.26 km link are presented.

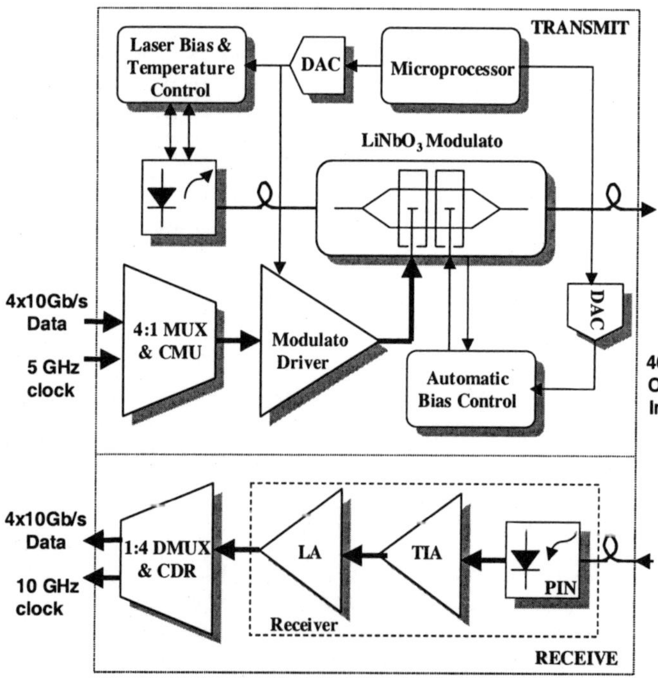

Fig. 1. Block diagram of the 40 Gb/s InP opto-electronic transceiver.

2. TDM System Configuration

Figure 1 shows the block diagram of the 40 Gb/s transceiver. The transmit section consists of a 4:1 MUX / CMU, a modulator driver, an optical modulator and a CW laser. The transmitter takes in four channels of parallel 10 Gb/s current mode logic level (CML) inputs and multiplexes it to a 40 Gb/s serial bit stream using the 4:1 MUX. A 20 GHz on-chip voltage controlled oscillator (VCO) in the MUX is divided down and locked to the 5 GHz reference clock input. The MUX provides a 700 mV_{pp} differential output into the modulator driver.

The modulator driver requires at least 20 dB gain, greater than 40 GHz bandwidth and large modulation swing. Lithium Niobate EO modulators require up to 6 V_{pp} drive and EA modulators could require up to 3 V_{pp} swing to obtain good extinction ratio. Reported distributed amplifier (DA) based drivers[3,4] have voltage gains typically less than 10 dB under large signal limiting conditions. Then pre-amplifier IC's are required between the MUX and the driver. Also off-chip broadband bias-tees are required to interface these different IC's with the penalty of increased size, complexity and packaging cost. Alternatively, high gain single chip lumped drivers employing cascaded differential amplifiers offer higher gain and differential output[5,6,7] but offer reduced bandwidth or lesser swing due to capacitances associated with the large transistor sizing at the output.

We use a single chip differential modulator driver based on a topology consisting of broadband lumped preamplifier stages and a differential DA stage. With 30 dB small signal gain and 41 GHz bandwidth, a 500 mV_{pp} differential input drive is sufficient to provide 6 V_{pp} differential output from a single chip and without external bias-tees. A second IC with similar performance, but with a single-ended output was also designed to drive EA modulators requiring 3 V_{pp} drive.

The differential driver was used to drive a differential lithium niobate EO modulator that modulates a 1550 nm CW laser (fig. 2). On-board micro-controller with 10–bit digital to analog converters (DAC) is used to control the laser temperature and bias, the driver swing and cross-point, and the modulator bias. An automatic bias control (ABC) loop maintains the modulator at its optimum bias point over temperature and lifetime, to provide the best extinction ratio.

Receiver sensitivity is one of the key parameters that decides repeater spacing in optical communication systems, and also imposes stringent performance requirements on optical amplifiers within the link. Therefore optical receivers influence the performance and cost of optical communication networks to a great extent. A number of wide-band optical receivers for 40 Gb/s systems have been reported[8,9,10] that uses optical amplifiers to amplify the signal before it undergoes an optical to electrical (O–E) conversion. These amplifiers can help to maintain a constant signal power incident on the photo-diode, which performs the O–E conversion. However they add to the system size, cost and complexity.

For our prototype we have developed a receiver with a PIN, TIA and LA dies integrated in a single package that achieves back to back sensitivity of –9.1 dBm

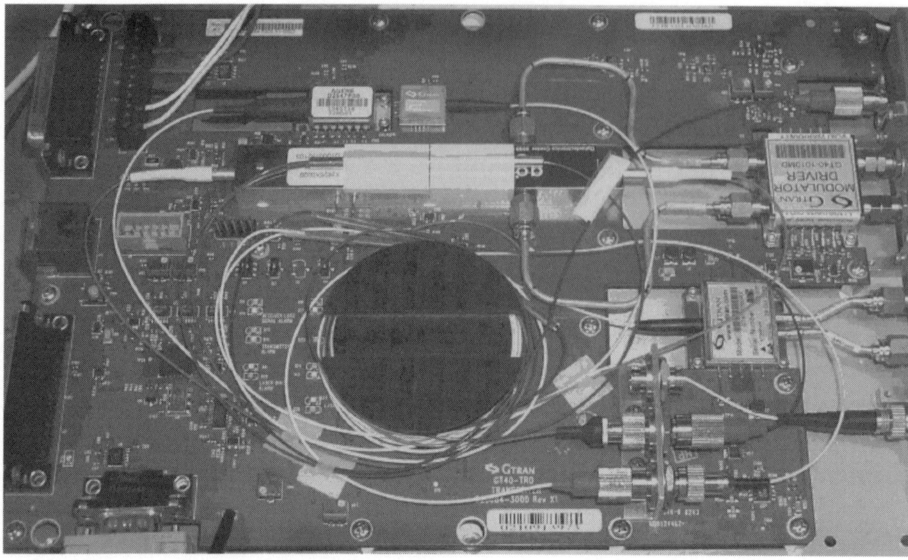

Fig. 2. Photograph of the 40 Gb/s transceiver prototype.

at a BER of 10^{-12}, as well as a wide dynamic range of 12.5 dB, measured using a 2^{31}–1 non-return to zero (NRZ) pseudo-random bit sequence (PRBS). The receiver provides a 900 mV_{pp} differential drive into the DEMUX. The 1:4 DEMUX IC has an integrated clock and data recovery circuit that synchronize the on-chip 40 GHz VCO to the phase of the incoming data. A 10 GHz clock, in phase with the demultiplexed output data is made available.

3. IC Technology

The high speed analog and digital ICs presented in this work (fig. 3)were designed in an InAlAs / InGaAs / InP DHBT technology[11]. The smallest devices with 1×3 mm^2 emitter area exhibit a typical current gain of 40, with a 140 GHz unity current gain cut-off frequency (f_τ) and 160 GHz power gain cut-off frequency (f_{max}). Carbon doping in the base improves reliability at high current density. Commercial 4 inch GaAs HBT wafer processing techniques were used. The process includes two layers of interconnect metal with polyimide as the dielectric, Si_3N_4 MIM capacitors with 0.36 fF/mm^2 capacitance, NiCr resistors with 50 Ω/□ resistivity and CPW transmission line wiring environment. The capabilities of the device technology are highlighted in the following circuits that also form critical blocks in the transceiver.

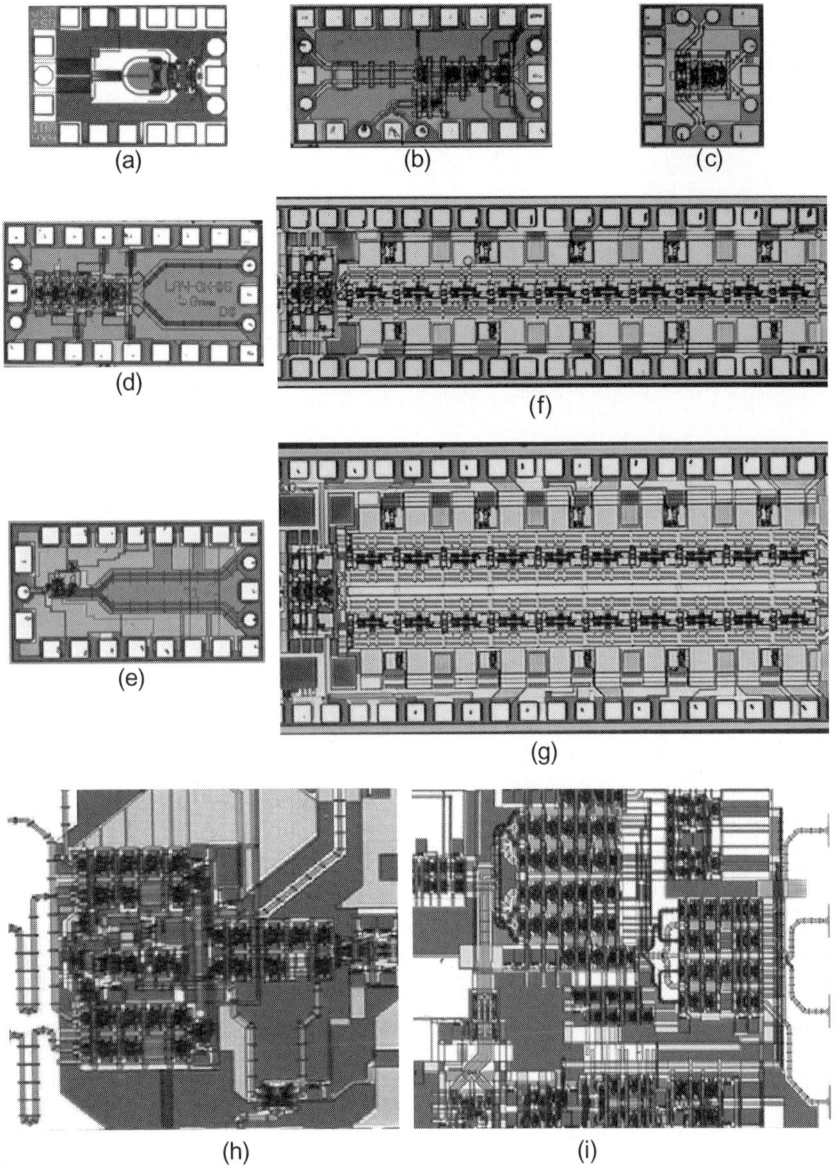

Fig. 3. Photograph of the InP HBT chip-set: (a) VCO, (b) D-flip-flop, (c) static divider, (d) LA, (e) TIA, (f) EA modulator driver, (g) EO modulator driver, (h) 4:1 MUX/CMU, and (i) 1:4 DEMUX/CDR.

3.1. *Voltage Controlled Oscillator*

A 40 GHz VCO forms a key component in the demultiplexer. A modified Colpitts architecture with a unique resonator was used to maximize the tuning range[12]. Simplified schematic of the VCO core is shown in fig. 4, where the transmission lines T_1 and T_2 provide the inductance, and diodes D_1–D_4 provide the capacitance for the differential Colpitts oscillator. The sizes of the varactors need to be maximized for largest tuning range and consequently the inductance of the transmission lines need to be reduced to keep the center oscillation frequency at 43 GHz. However this will result in reduction of Q-factor and loop gain. Hence the size of the varactors is eventually limited by minimum loop gain and Q-factor demanded by phase noise specification.

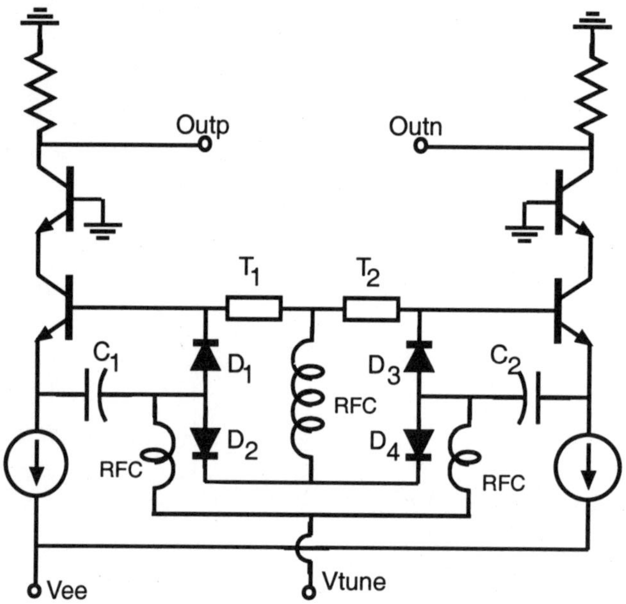

Fig. 4. Simplified schematic of the differential Colpitts VCO with cascode output stages.

In the layout, T_1 and T_2 were arranged into U-shaped tanks with opening at one end to accommodate transistors and varactors in the loop and form a unique resonator, where the impacts of interconnects in the tank circuit are minimized. 200Ω resistor and high impedance lines were used to implement the RF Choke (RFC).

The IC uses a single −4.2 V supply and consumes 50 mW. Including both buffer and pad frame, the VCO measures only 700 μm^2 (fig. 3(a)). A tuning range of 37–50 GHz (30%) is obtained over a tuning voltage of 1.2 V. This allows a single

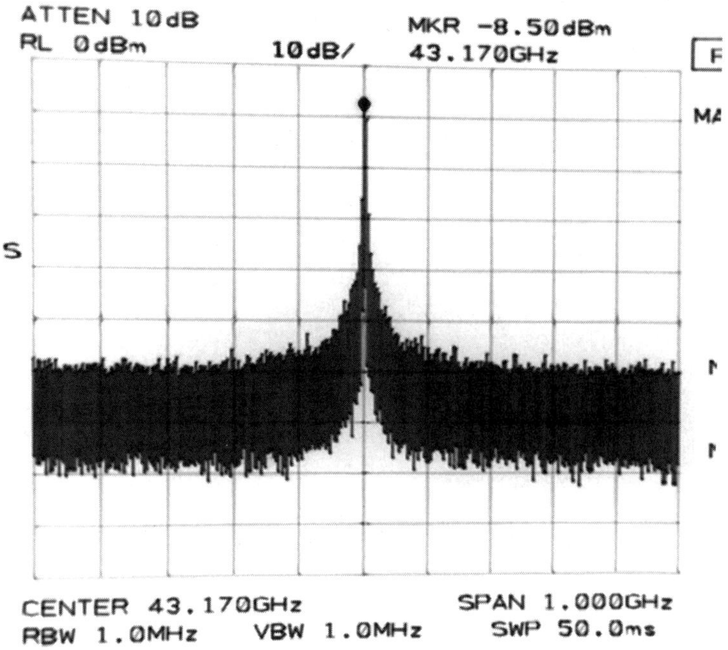

Fig. 5. Measured 43 GHz spectrum of the VCO.

multiplexer to cover all OC-768 data rates, from the basic rate (40 Gb/s) to strong-FEC rate (50 Gb/s). The VCO delivers –7 dBm output to a 100 Ω differential off-chip load through a buffer, with excellent flatness over the entire tuning range. The measured spectrum of the VCO operating at 43 GHz (fig. 5) shows –8.5 dBm single-ended output power without accounting for cable losses. Phase noise is measured to be ~100 dBc at 10 MHz offset.

3.2. *D-flip-flop*

High-speed master-slave D-flip-flop (MS-DFF) is a critical component in various circuits used for optical communication systems. Demultiplexers can be built by setting the clock frequency equal to half the data bit rate[13]. Multiplexers with full-rate retiming use them for data reshaping to improve the transmitted signal quality[14].

The flip-flop IC (fig. 3(b)) consists of data and clock input buffers, MS-DFF core and an output buffer. The input buffers were designed for 100 mV_{pp} sensitivity and the CML output buffer was designed for a 1 V_{pp} differential output swing. Emitter coupled logic was used in the MS-DFF core (fig. 6) to obtain high operating speed[15]. Fully symmetric differential layout was used within the individual cells. Further, all critical transmission lines were modeled and transmission lines between stages were

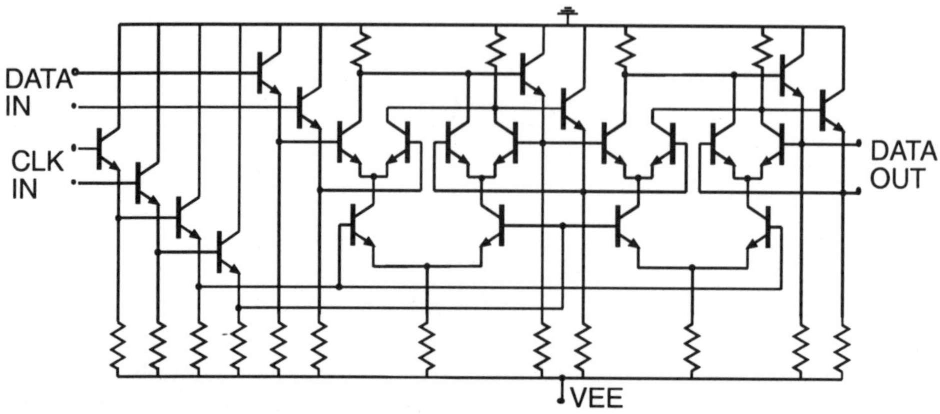

Fig. 6. Schematic of the high-speed master-slave D-flip-flop.

terminated at either end to minimize ringing. The IC measures 1.4×0.8 mm^2 and operates from a –4.2 V supply consuming 0.7 W. The DFF was used to retime a 40 Gb/s PRBS data. Error free operation was observed with a clock phase margin of 190°.

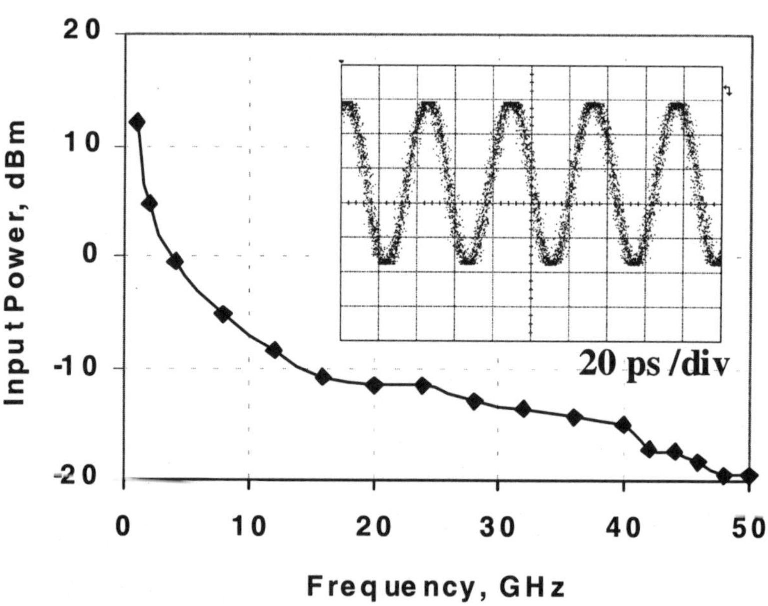

Fig. 7. Measured input sensitivity of the 2:1 static frequency divider.

3.3. *2:1 Static Frequency Divider*

The phase detector for the DEMUX requires a 20 GHz four-phase clock, which can accurately be generated by dividing the 40 GHz VCO output by a 2:1 frequency divider to obtain differential 0° and 90° phase-shifted clock signals.

A 2:1 static frequency divider was implemented from the above MS-DFF by feeding back the inverted slave output to the master input. Usually a dynamic frequency divider is used to obtain higher frequency performance but has an inherent drawback because of the lower operating frequency limit. However, the performance of a 2:1 static divider (fig. 3(c)) in this device technology is sufficient up to 50 GHz as seen from the input sensitivity plot (fig. 7). The inset shows the divided 25 GHz output for a 50 GHz input. A differential voltage output of 0.6 V_{pp} is obtained with 10 ps rise / fall times. The divider uses a single –4.2 V supply and consumes 0.55 W.

3.4. *Trans-impedance Amplifier*

A block diagram of the TIA (fig. 8) shows a linear trans-impedance stage followed by a gain stage that also converts the signal from single-ended to differential.

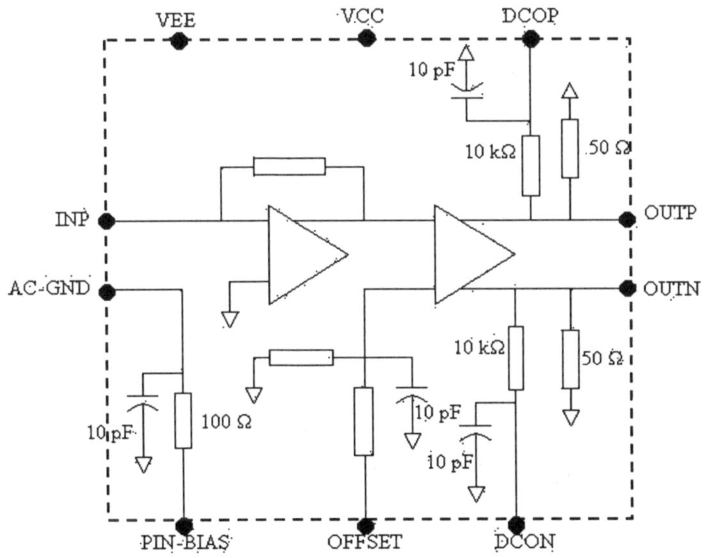

Fig. 8. Schematic block diagram of the TIA.

The trans-impedance stage of the TIA was optimized for bandwidth taking into account the photo-diode capacitance / resistance and the bond wire inductance.

At 40 Gb/s data rates, the parasitic bond wire inductances between the photodiode and TIA as well as the ground inductance associated with the bond wires connecting the TIA IC to the receiver package are critical and were accounted for in the design[16]. The differential design provides high power supply rejection. Additionally, the second output could be used for monitor purposes to measure the extinction ratio of the input optical signal. The single-ended to differential converter requires a reference signal that is dependent on the magnitude of the input photo-signal to balance the TIA outputs.

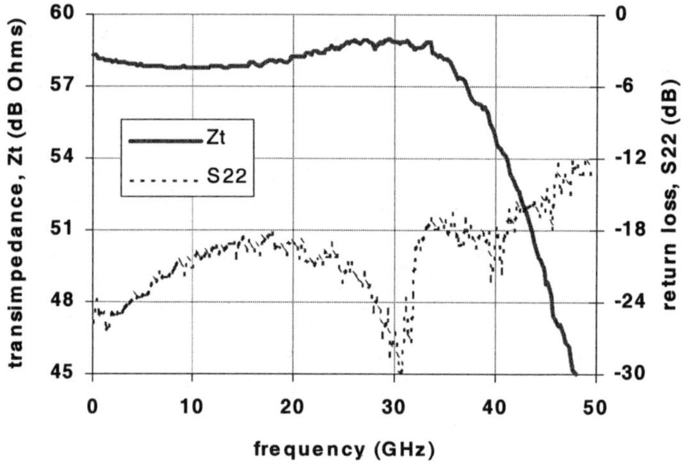

(a)

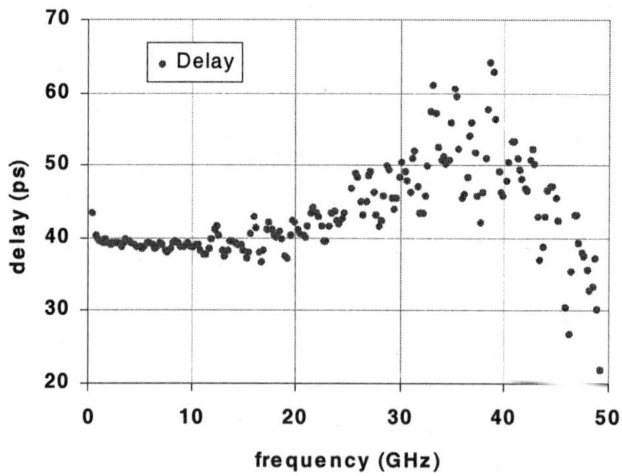

(b)

Fig. 9. Measured (a) trans-impedance, output match and (b) group delay response of the TIA.

The TIA die measures 1.4×0.8 mm^2 (fig. 3(e)). The IC uses +3.3 V and –5 V supply and consumes 0.4 W. A differential trans-impedance of 750 Ω with 37 GHz bandwidth and ±10 ps group delay variation is obtained (fig. 9).

3.5. *Limiting Amplifier*

To obtain high linear gain the LA uses alternating differential trans-admittance (TAS) and trans-impedance (TIS) stages (fig. 10). Low inter-stage impedance provided by this architecture[17] greatly extends the bandwidth. The four stage LA has CML interfaces with 50 Ω terminations, input threshold adjust and output offset adjust. The LA die measures 1.4×0.8 mm^2 (fig. 3(d)). The IC uses a single –5 V supply and consumes 0.7 W. The LA has 20 mV differential input sensitivity, 24 dB gain, 35 GHz bandwidth and provides 0.9 V_{pp} differential swing with 8.2 ps rise time, 6.4 ps fall time and 0.6 ps added root mean square (rms) jitter (fig. 11).

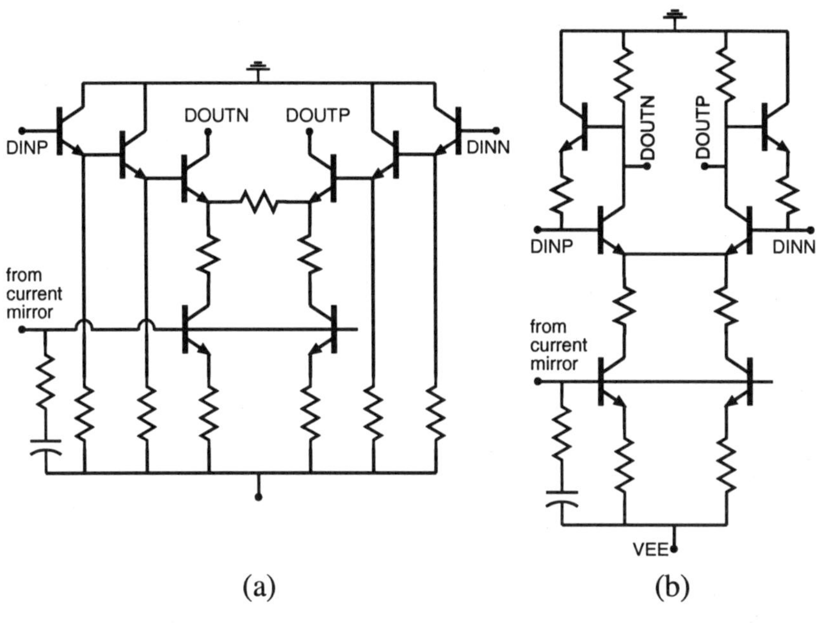

Fig. 10. Schematic of limiting amplifier (a) trans-admittance and (b) trans-impedance stages.

3.6. *EA Modulator Driver*

Figure 12 shows the schematic block diagram of the single-ended electro-absorption modulator driver. Simultaneous high gain, broad bandwidth and large swing was obtained in a single IC by using a lumped preamplifier stage to drive a 10–cell distributed output stage designed for 3 V_{pp} swing. The preamplifier consists of TIS-TAS stages similar to the limiting amplifier discussed earlier.

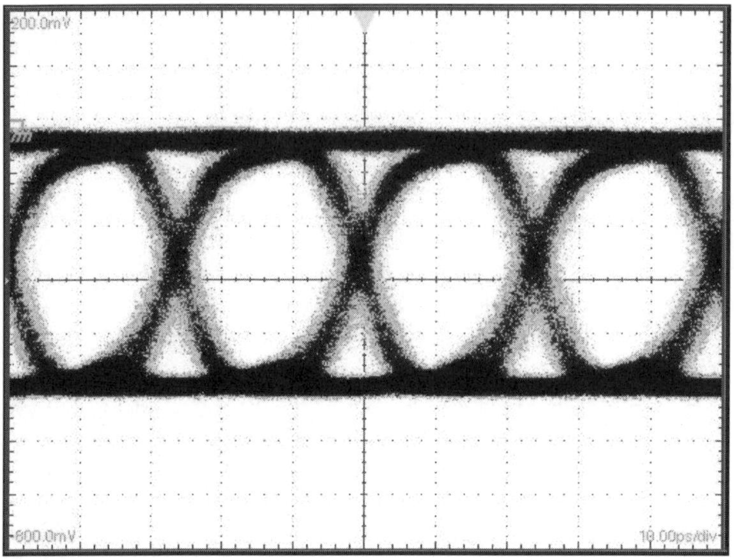

Fig. 11. Measured 40 Gb/s output from the LA.

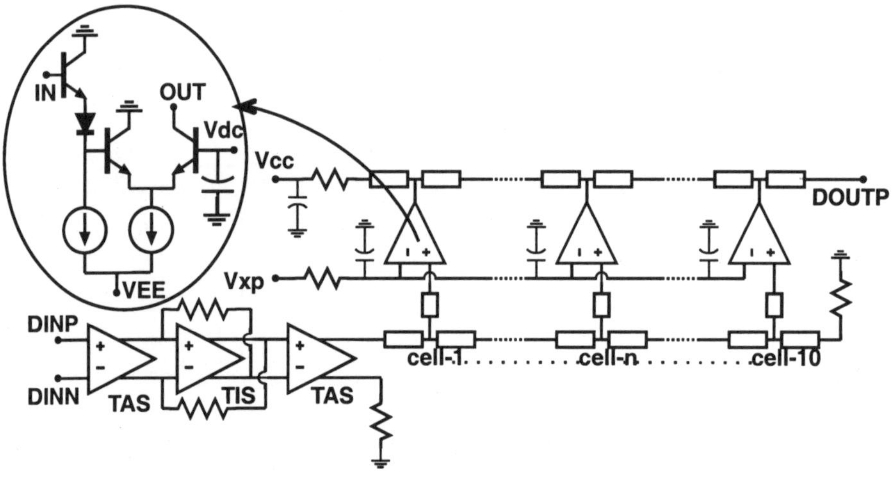

Fig. 12. Schematic block diagram of the single-ended modulator driver.

The cells of the distributed amplifier are made of emitter followers driving a quasi-differential pair (fig. 12). When driven single-ended they act as a common collector - common base pair (CC–CB). The CB device provides RC degeneration for the CC device. This reduces the capacitance seen at the base of the CC device by half, improving the bandwidth. The emitter-follower further decreases the capacitive loading on the input transmission line while providing level-shifting. The CC–CB pair also mitigates the miller effect just like the cascode pair, but without decreasing the voltage head-room available for the output transistor. Additionally unlike cascode pair, this topology offers easy biasing through current sources. The unused input of the quasi-differential pair is connected to a DC reference (V_{xp} in fig. 12) that sets the cross-point of the output.

The EA modulator driver ICs measure 3.3×1 mm^2 (fig. 3(f)) and uses +3.3 V and −4.2 V supplies consuming 1.8 W when driving full modulation swing. Measured single-ended output at 40 Gb/s (fig. 13) shows 3.0 V_{pp} swing with 20–40% rise and fall times of 8.4 ps and 7.2 ps respectively. Measured RMS jitter of 1.2 ps corresponds to an added jitter of 0.5 ps.

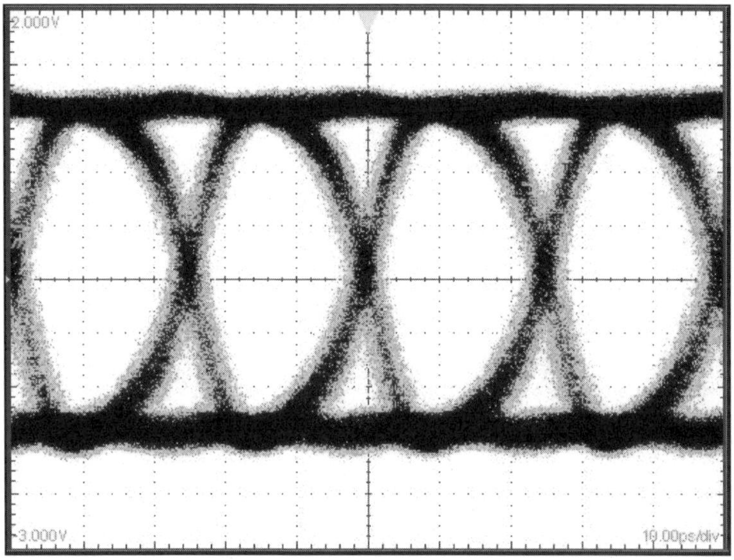

Fig. 13. Measured 40 Gb/s output from the modulator driver.

3.7. *EO Modulator Driver*

A second driver with 6 V_{pp} differential output, intended for differential EO modulators, was implemented by using a similar distributed amplifier at the inverting output of the preamplifier[18].

The distributed amplifier was optimized for best large signal performance. As a starting point, frequency domain simulations using linear models were used to design the transmission line sections for maximum overall bandwidth and minimum return losses. In large signal operation, variation of the junction capacitances have to be accounted. Time-domain simulation using a step input was used to further optimize the transmission line lengths to minimize reflections. Also the loaded transmission line propagation delays were matched for the input and output lines. This approach optimizes the design for reduced jitter, faster transition times and minimum overshoots.

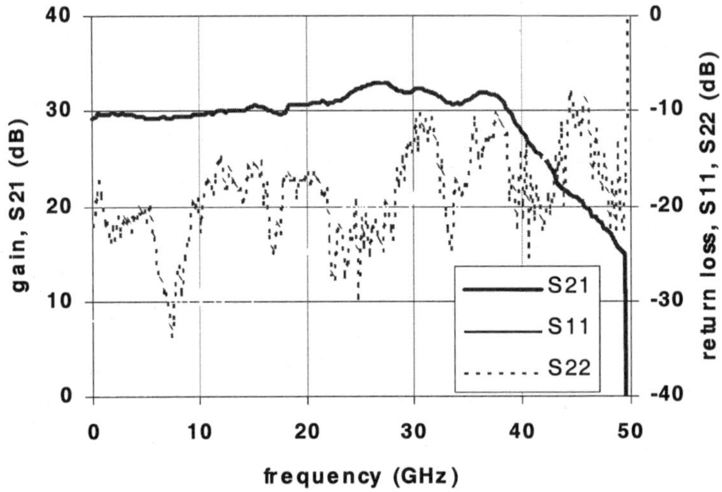

Fig. 14. On-wafer S-parameters of the modulator driver.

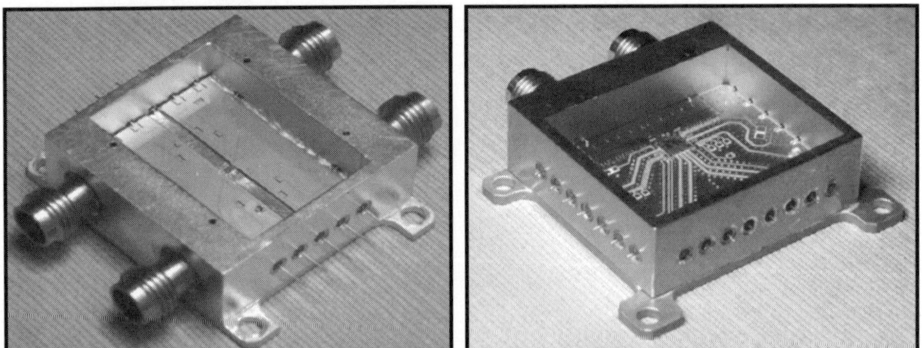

Fig. 15. Photograph of the (a) modulator driver and (b) 4:1 MUX / CMU packages.

Simulations showed that a positive small-signal gain slope with frequency was required for faster transition times under large signal operation. The distributed amplifier was designed for 3 dB small signal gain peaking about 40 GHz. Measured S-parameters of the modulator driver (fig. 14) shows a mid-band gain of 30 dB, 41 GHz bandwidth, and lower than 10 dB return losses up to 43 GHz.

The differential driver IC measures 3.3×1.5 mm^2 (fig. 3(g)) and was packaged in a high frequency ceramic package with 2.4 mm connectors (fig. 15(a)). These drivers also operate from +3.3 V and -4.2 V supplies, and consumes 2.8 W when driving full modulation swing. Additional features like modulation current control to adjust the output swing, cross-point control to change the zero crossing from 30% to 80% (fig. 16) and output power down option were also implemented. Monitor pin for modulation current is available for external closed-loop automatic power control.

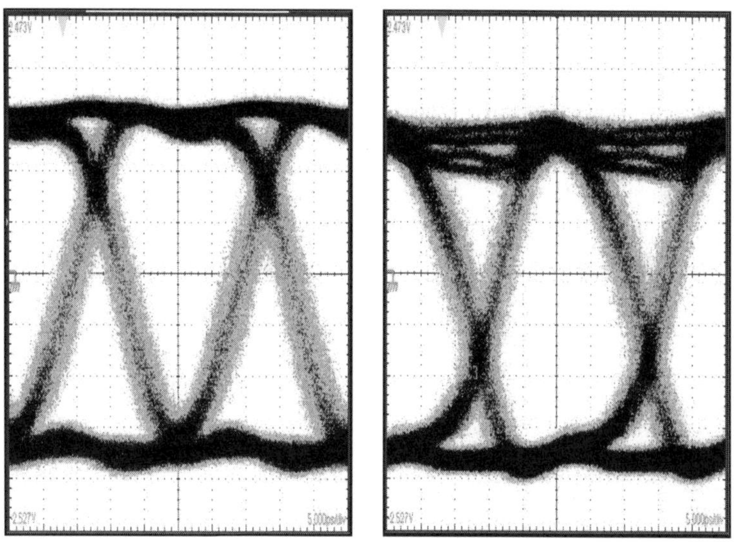

Fig. 16. Eye crossing control in the modulator driver : (a) 80% crossing and (b) 30% crossing.

3.8. *Multiplexer / Clock Multiplication Unit*

A block diagram of the MUX / CMU is shown in fig. 17. The MUX core uses a pipelined binary tree structure. The 2:1 selectors use ECL latches with additional buffers to regenerate the data. The CMU phase locked loop consists of a phase detector, a proportional-integral loop filter, a 20 GHz VCO and static frequency dividers. The VCO is phase-frequency locked to the 5 GHz reference input via the static dividers and the internal loop filter. The phase locked loop has a locking range of 36–45 GHz and requires only a single external capacitor. The clock distribution circuit delivers clock signals with proper phase and frequency for all the latches. Differential CML logic levels with 50 Ω terminations was used for the in-

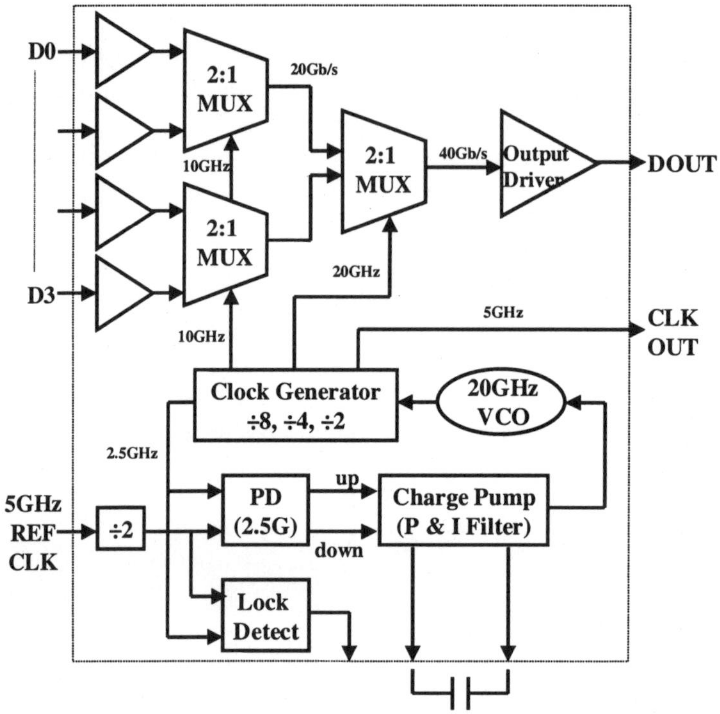

Fig. 17. Schematic block diagram of the 4:1 MUX / CMU.

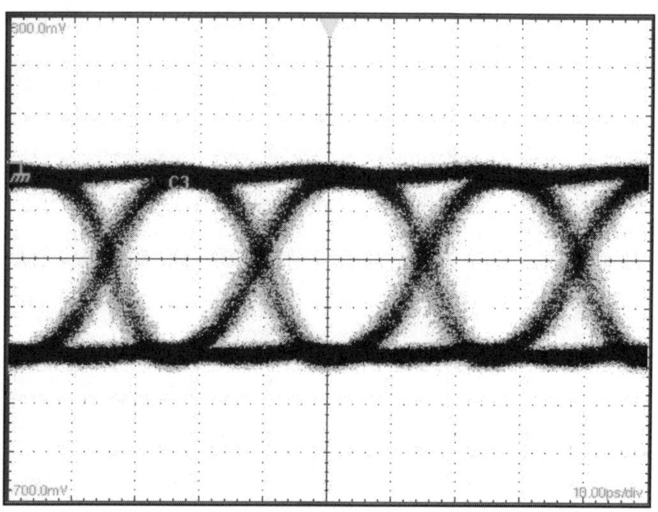

Fig. 18. Measured 40 Gb/s output from the 4:1 MUX.

put and output interfaces. The output driver was designed to provide a variable 600–900 mV_{pp} differential swing.

The IC (fig. 3(h)) measures 3.3×3.3 mm^2 and operates of a single –4.2 V supply, consuming 2.7 W. The MUX was packaged in a ceramic package with 2.4 mm connectors and tested (fig. 15(b)). Measured 40 Gb/s single-ended output (fig. 18) shows 360 mV swing with rise / fall times of 9.0 ps / 7.2 ps and 1.1 ps rms jitter using a 50 GHz sampling module.

3.9. *Demultiplexer / Clock and Data Recovery*

The 1:4 DEMUX (fig. 19) is based on a half-bit-rate architecture that has been demonstrated earlier for 40 Gb/s application using SiGe HBT technology[19]. It incorporates an Alexander phase detector[20] for clock and data recovery. On-chip 40 GHz VCO is synchronized to the incoming data using a phase-locked loop consisting of a phase-frequency detector and a proportional-integral filter. The IC requires a 622 MHz reference clock input and outputs a differential 10 GHz recovered clock. The chip also includes a lock detect circuitry and requires only two external components, a loop filter capacitor, and a lock reference resistor.

The IC (fig. 3(i)) measures 3.3×3.3 mm^2, uses a single –4.2 V supply and con-

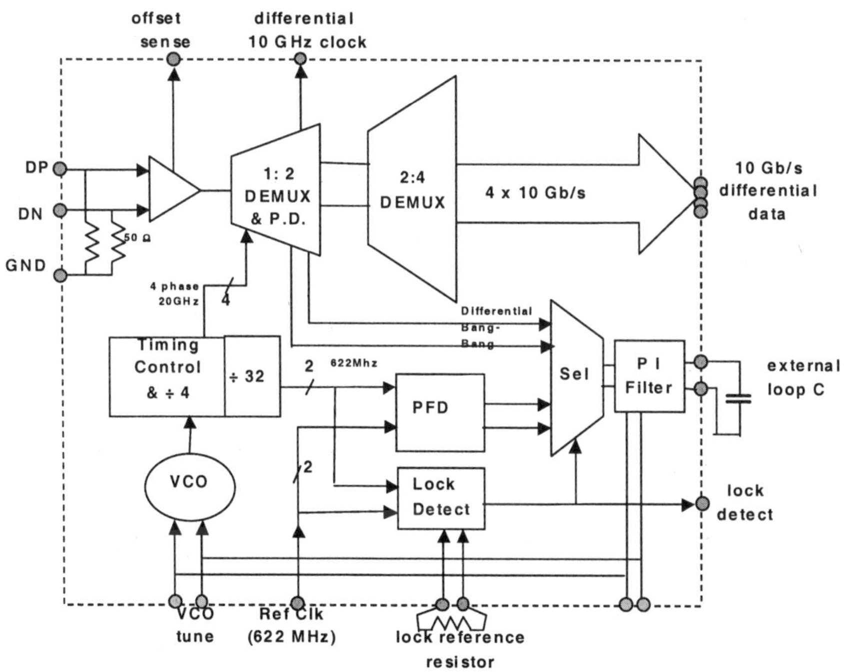

Fig. 19. Schematic block diagram of the 1:4 DEMUX / CDR.

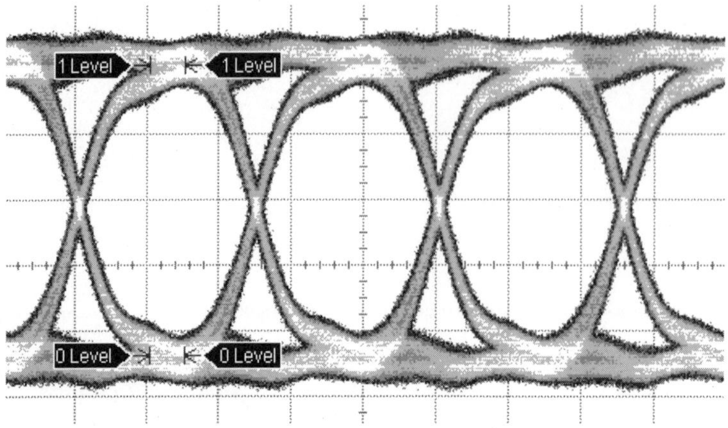

Fig. 20. Recovered 10 Gb/s data from the 1:4 DEMUX / CDR.

sumes 3.5 W. A similar package as the MUX was used. Figure 20 shows one channel of the recovered 10 Gb/s data from a 40 Gb/s bit pattern.

4. Transceiver Performance

The 40 GB/s TDM transceiver described in section 2 was implemented using the components described in the last section. Measured performance is detailed below.

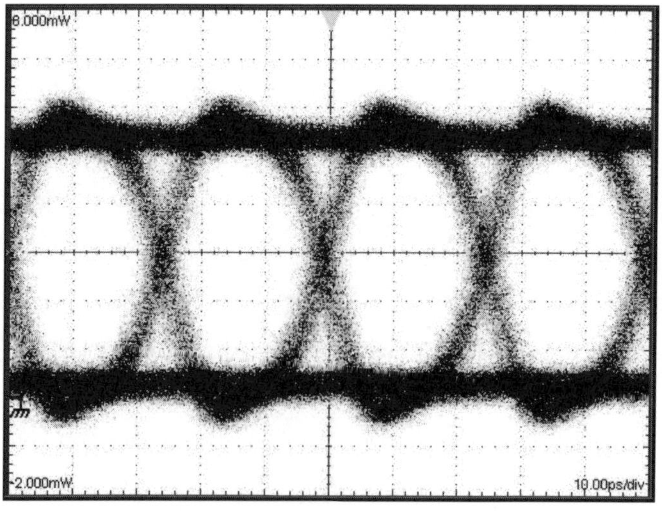

Fig. 21. Measured 40 Gb/s optical output from the Lithium Niobate EO modulator.

4.1. *Transmit*

4 channels of de-correlated 10 Gb/s 2^{31}–1 NRZ PRBS was used to test the transmitter. The 1550 nm CW laser source was biased to provide 13 dBm output power into the lithium niobate modulator. The modulator was driven with 5.6V_{pp} from the differential driver. The measured optical output (fig. 21) exhibits 4.4 dBm power with 12.9 dB extinction ratio and 1.4 ps rms jitter, corresponding to 0.9 ps of added rms jitter.

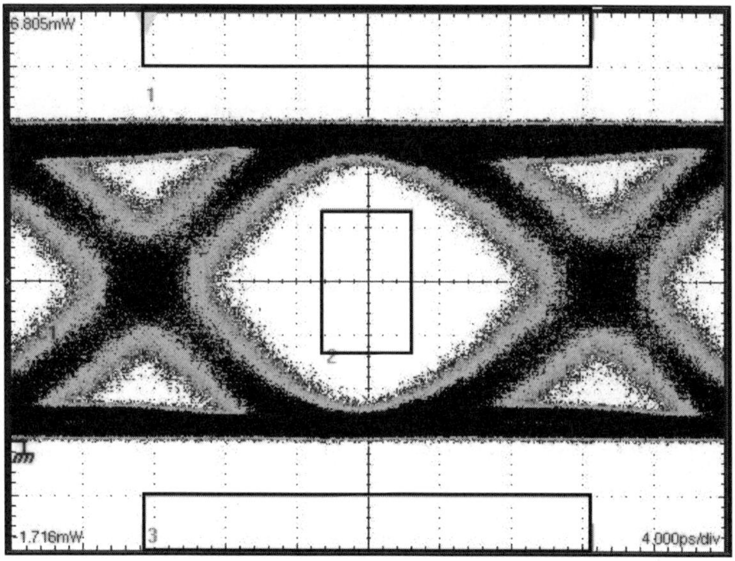

Fig. 22. OC-768 eye mask test on the transmitted optical output.

OC–768 mask test performed with a 50 GHz optical head shows sufficient margin for error-free performance (fig. 22). After 2.26 km transmission over single mode fiber (SMF) the extinction ratio dropped to 10.3 dB, with 1.6 ps rms jitter and 40% crossing.

A second transceiver implemented with the single-ended driver driving an EA modulator achieved 2.4 dBm optical output with 10.9 dB extinction ratio and 1.5 ps rms jitter (fig. 23).

4.2. *Receive*

The integrated receiver package[21] containing the PIN, TIA and LA incorporates a DC restoration loop (fig. 24). This provides a reference signal for the single-ended to differential converter on the TIA. This loop was designed so that the low frequency components attenuated by the loop is a small fraction of the total energy

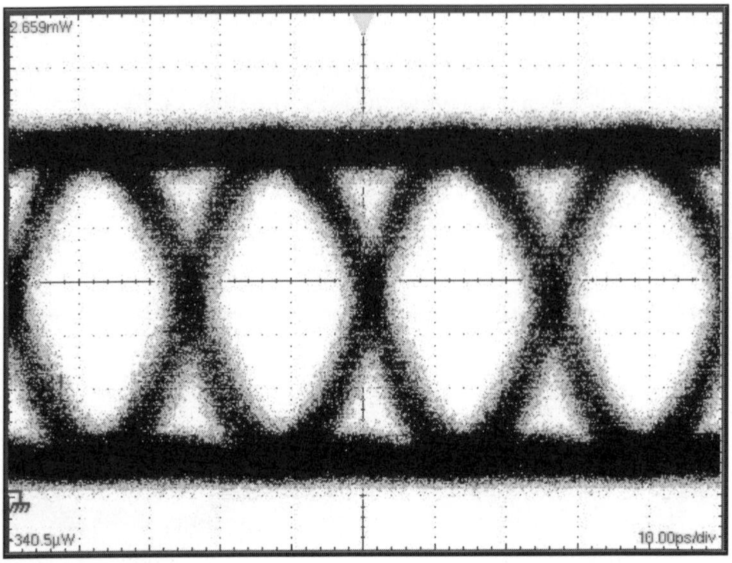

Fig. 23. Measured 40 Gb/s optical output from the EA modulator.

present in the input signal. The receiver uses +5 V, +3.3 V and –5 V supplies and consumes 1.1 W. The photo-diode used has a responsivity of 0.76 A/W. The differential conversion gain of the receiver module including the PIN, TIA and LA is about 8000 V/W, and the small signal bandwidth is 35 GHz.

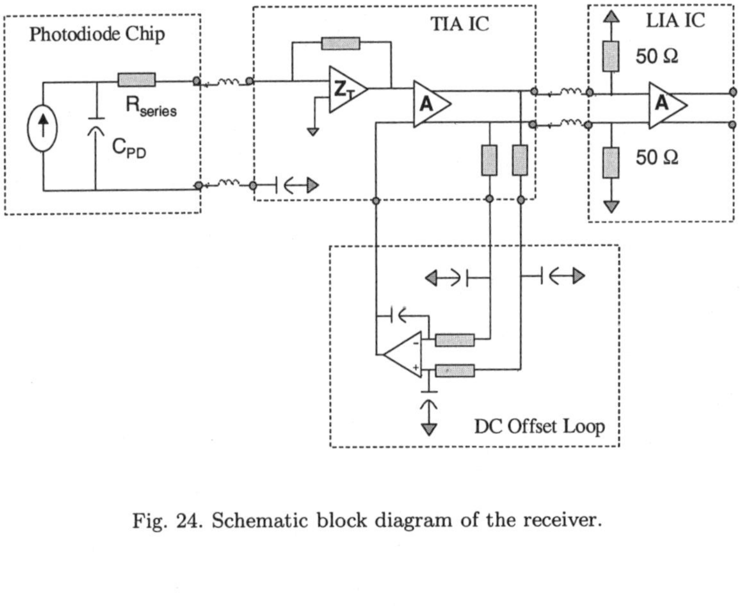

Fig. 24. Schematic block diagram of the receiver.

BER measurements were performed using the transmitter 40 Gb/s optical output. The received signal was demultiplexed into 4 channels of 10 Gb/s signal and the bit error rate was tested. A plot of BER vs. the input optical power into the receiver after 0 km and 2.26 km of SMF fiber is shown in fig. 25. A sensitivity of –9.1 dBm at a BER of 10^{-12} and an overload margin of 3.4 dBm was obtained corresponding to a dynamic range of 12.5 dBm. No dependency on pattern length was observed indicating that the gain and group delay performance at the low frequency cut-off of the receiver was adequate. A dispersion penalty of 0.24 dB was observed after transmission over 2.26 km of SMF.

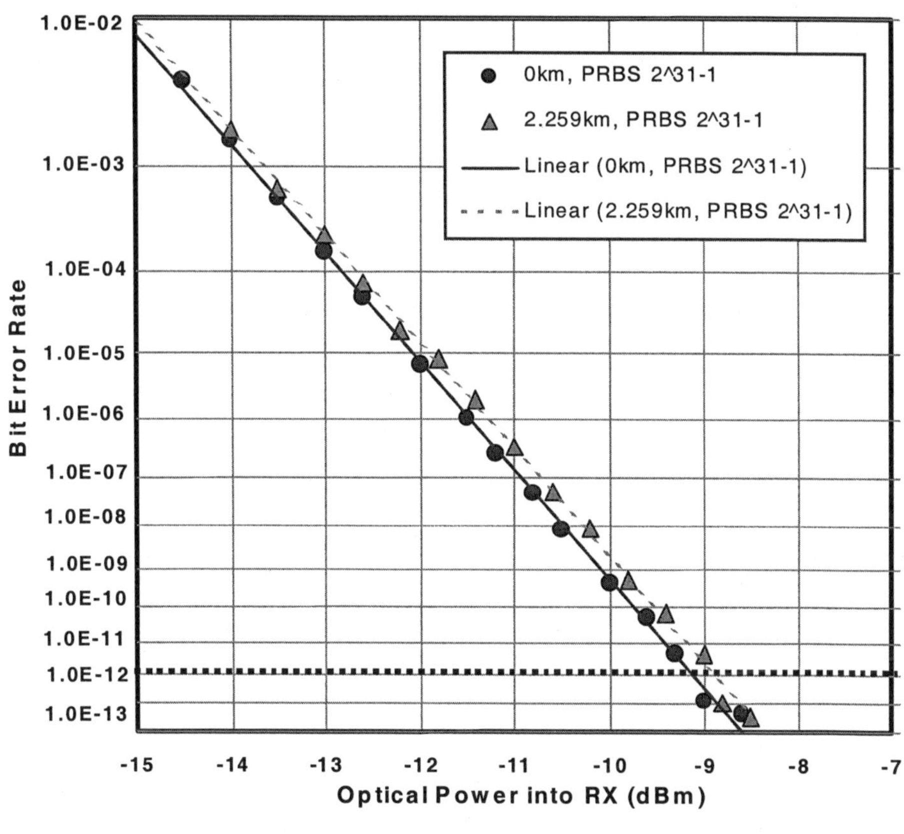

Fig. 25. Measured BER performance of the 40 Gb/s receiver.

5. Conclusion

We have demonstrated a TDM system transceiver prototype for 40 Gb/s NRZ transmission. The transmit unit multiplexes 4 channels of 10 Gb/s CML level inputs into a 40 Gb/s optical output with 4.4 dBm average power, 12.9 dB extinction ratio and 0.9 ps added rms jitter. The receive unit demultiplexed the 40 Gb/s optical signal into 4 channels of 10 Gb/s CML level data with a receive sensitivity of −9.1 dBm (at a BER of 10^{-12} for a 2^{31}–1 PRBS) and a dynamic range of 12.5 dB without the use of optical amplifiers. A dispersion penalty of 0.24 dB was observed for a 2.26 km link. The total power consumption for the transmitter and receiver was 13.3 W. To achieve this performance we used packaged IC's developed in InP DHBT technology. These IC's provide sufficient performance and level of integration for use in test equipments and communication systems. Further improvement in packaging and integration level is required to meet cost and size constraints in transponders.

Acknowledgments

The authors would like to thank Chris Winczewski, Sang Vu and Nga Tran for the meticulous IC layout work. The authors are indebted to Dino Mensa for the transistor design and device modeling, Jimmy Chow for test and measurement, Liyang Zhang for VCO circuit design, Ram Panicker and Ivair Gontijo for the package design, Mark Rubin for the printed circuit board design, Robert Franks for the mechanical design, and to Erik Loose for the software programming. The painstaking work of wire bonding and assembly by Angelica Rodriguez, Freddie Flores and Raul Fernandez is gratefully acknowledged.

References

1. T. Otsuji, Y. Imai, E. Sano, S. Kimura, S. Yamaguchi, M. Yoneyama, T. Enoki, Y. Umeda, "40-Gb/s ICs for future lightwave communications systems", *IEEE J. Solid-State Circ.* **32** (Sept. 1997) 1363–1370.
2. H.-M. Rein, "Si and SiGe bipolar ICs for 10 to 40 Gb/s optical-fiber TDM links", *Int. J. High Speed Elec. and Sys.* **9** (June 1998) 347-383.
3. J.M. Carroll, M. Coutant, M.S. Heins, C.F. Campbell, E. Reese, "0.25mm PHEMT 40Gb/s E/O modulator drivers", *2002 IEEE MTT-S Int. Microwave Symp. Dig.* (June 2002) 489–492.
4. M.S. Heins, J.M. Carroll, M.-Y. Kao, C. F. Steinbeiser, T. R. Landon, C. F. Campbell, "An ultra-wideband GaAs PHEMT driver amplifier for fiber optic communications at 40 Gb/s and beyond", *Optical Fiber Comm. Conf. Tech. Dig.* (March 2002) 273–274.
5. H. Suzuki, K. Watanabe, K. Ishikawa, H. Masuda, K. Ouchi, T. Tanoue, R. Takeyari, "Very-high-speed InP/InGaAs HBT IC's for optical transmission systems", *IEEE J. Solid-State Circ.* **33** (Sept. 1998) 1313–1319.
6. Z. Lao, A. Thiede, U. Nowotny, H. Lienhart, V. Hurm, M. Schlechtweg, J. Hornung, W. Bronner, K. Kohler, A. Hulsmann, B. Raynor, T. Jakobus, "40-Gb/s High-power modulator driver IC for lightwave communication systems", *IEEE J. Solid-State Circ.* **33** (Sept. 1998) 1520–1526.
7. M. Meghelli, M. Bouche, A. Konczykowska, "High power and high speed InP DHBT driver IC's for laser modulation", *IEEE J. Solid-State Circ.* **33** (Sept. 1998) 1411–1416.
8. S. Kuwahara, K. Yonenaga, Y. Miyamoto, K. Murata, H. Kitabayashi, N. Himizu, H. Toba, "43 Gbit/s, 200 km unrepeatered transmission experiment using high sensitivity digital OEIC receiver module", *Electron. Lett.* **38** (Jan. 2002) 38–40.

9. R. Ohira, Y. Amamiya, T. Niwa, N. Nagano, T. Takeuchi, C. Kurioka, T. Chuzenji, K. Fukuchi, "A high sensitivity 40 Gbit/s optical receiver using packaged GaAs HBT-ICs", *Optical Fiber Comm. Conf. Tech. Dig.* (1998) Paper WI4.
10. K. Takahata, Y. Miyamoto, Y. Muramoto, H. Fukano, Y. Matsuoka, "Ultrafast monolithic receiver OEIC module operating at over 40Gbit/s", *Electron. Lett.* **35** Jan. 1999 322–324.
11. K. Krishnamurthy, R. Vetury, J. Xu, A. Shou, S. Jaganathan, K. Cheng, J. Chow, D. Mensa, L. Zhang, I. Gontijo, S. Vu, C. Winczewski, Y.-Z. Liu, R. Pullela, M. Rodwell, "40 Gb/s TDM system using InP HBT IC technology", *2003 IEEE MTT-S Int. Microwave Symp. Dig.* **2** (8–13 June 2003) 1189–1192.
12. L. Zhang, R. Pullela, C. Winczewski, J. Chow, D. Mensa, S. Jaganathan, M. Rodwell, R. Yu, "A 37 50 GHz InP HBT VCO IC for OC-768 fiber optic communication applications", *2002 IEEE RFIC Symp. Dig.* (June 2002) 85–88.
13. A. Felder, M. Moller, J. Popp, J. Bock and H–M. Rein, "46 Gb/s DEMUX, 50 Gb/s MUX, and Static Frequency Divider in Silicon Bipolar Technology", *IEEE J. Solid-State Circ.* **31** (Apr. 1996) 481–486.
14. Y. Nakasha, T. Suzuki, H. Kano, K. Tsukashima, A. Ohya, K. Sawada, K. Makiyama, T. Takahashi, M. Nishi, T. Hirose, M. Takikawa and Y. Watanabe, "A 43–Gb/s Full-Rate-Clock 4:1 Multiplexer in InP-Based HEMT Technology", *IEEE J. Solid-State Circ.* **37** (Dec. 2002) 1703–1709.
15. K. Krishnamurthy, J. Chow, D. Mensa and R. Pullela, "43 Gb/s Decision Circuits in InP DHBT Technology", *IEEE Microwave and Wireless Components Lett.* **14** (Jan. 2004) 28–30.
16. R. Vetury, I. Gontijo. Y.-Z. Liu, K. Krishnamurthy, R. Pullela, M. J. W. Rodwell, "High sensitivity and wide-dynamic-range optical receivers for 40 Gbit/s optical communication networks", *Electron. Lett.* **39** (Jan. 2003) 91–92.
17. H.-M. Rein, "Design considerations for very-high-speed Si-bipolar IC's operating up to 50 Gb/s", *IEEE J. Solid-State Circ.* **31** (Aug. 1996) 1076–1090.
18. K. Krishnamurthy, R. Pullela, J. Chow, J. Xu, S. Jaganathan, D. Mensa, M. Rodwell, "High gain 40 Gb/s InP HBT drivers for EO/EA modulators", *Optical Fiber Comm. Conf. Tech. Dig.* **2** (Mar. 2003) 761–762.
19. M. Reinhold, C. Dorschky, E. Rose, R. Pullela, P. Mayer, F. Kunz, Y. Baeyens, T. Link, J-P. Mattia, "A fully integrated 40 Gb/s clock and data recovery IC with 1:4 DEMUX in SiGe technology", *IEEE J. Solid-State Circ.* **36** (Dec. 2001) 1937–1945.
20. J. J. D. H. Alexander, ']lq Clock recovery from random binary signals", *Electron. Lett.* **11** (1975) 541–542.
21. K. Krishnamurthy, R. Vetury, Y.-Z. Liu, M. J. W. Rodwell, R. Pullela, "40 Gbit/s optical receiver module with high conversion gain and sensitivity", *Electron. Lett.* **39** (Nov. 2003) 1738–1739.

International Journal of High Speed Electronics and Systems
Vol. 15, No. 3 (2005) 667–704

ENHANCED NETWORK SIGNALING FOR 10 GIGABIT ETHERNET TO ACHIEVE A LAN-WAN SEAMLESS INTERFACE AND ITS IMPLEMENTATION IN THE PHY-LSI/TRANSCEIVER MODULE

Haruhiko Ichino[†], Kazuhiko Terada[†], Kenji Kawai[*], Osamu Ishida[$],
Keishi Kishine[†], Noboru Iwasaki[†]

[†]*NTT Microsystem Integration Labratories, NTT Corp.,*
3-1 Morinosato Wakamiya, Atsugi, Kanagawa 243-0198, JAPAN
[*] *NTT Cyber Solutions Laboratories, NTT Corp.,*
[$] *NTT Network Innovation Laboratories, NTT Corp.,*
1-1 Hikarinooka, Yokosuka, Kanagawa 239-0847, JAPAN
[†]*ichino@aecl.ntt.co.jp, kterada@aecl.ntt.co.jp,*
kenji.kawai@aecl.ntt.co.jp, osamu.ishida@lab.ntt.co.jp,
kishine@aecl.ntt.co.jp, niwasaki@aecl.ntt.co.jp

Tetsuhiro Fukao[‡], Shinji Shibao[‡], Kazuo Kubo[§], Harufusa Kondoh[‡]

[‡]*High-Frequency and Optical Device Works, Mitsubishi Electric Corp.,*
4-1 Mizuhara, Itami, Hyogo 664-8641, JAPAN
[§]*Information Technology R&D center, Mitsubishi Electric Corp.*
5-1-1, Ofuna, Kamakura, Kanagawa 247-8501, JAPAN
fukao@lsi.melco.co.jp, shibao.shinji@lsi.melco.co.jp,
kkubo@isl.melco.co.jp, kondoh@lsi.melco.co.jp

This paper describes Inter-frame Ling Signaling (ILS) protocol technology and its implementation in a 10-Gigabit Ethernet (10GbE) physical layer (PHY) LSI and optical transceiver module. The ILS can realize SDH/SONET-compatible operations, administration, maintenance, and provisioning (OAM&P) functions for PHY of 10GbE networks by inserting OAM&P information in the inter-frame gap periods. The coding features include 10GbE PHY upper compatibility, error detection capability, and disparity neutral characteristics. The 10GbE PHY LSI with the ILS function was fabricated using 0.18-um SOI CMOS technology to achieve high-speed and low-power performance. The main features of the high-speed circuit design are an inductive-load buffer, high-speed clock and data recovery (CDR), and pre-emphasis/equalizer circuits. The LSI function meets IEEE 802.3ae specifications with the very low-power consumption of about 1.2W. An XENPAK transceiver with ILS function using the LSI was also fabricated to make it easy to implement the ILS function in optical transport network systems such as WDM, OADM, OXC systems and L2/L3 switches. The key features of the module include a 1.3-um DFB laser module with highly reliable transmission performance and a small EML driver integrated module. ILS protocol operation is verified by experiment with XENPAK modules emulating the optical transport system.

Keywords: OAM&P; 10G Ethernet; PHY LSI.

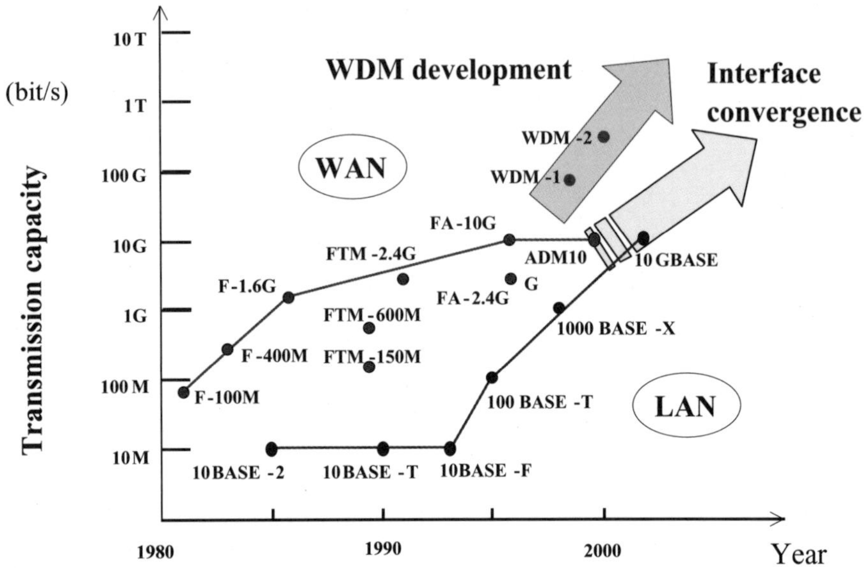

Fig. 1. Trend in LAN and WAN interface speeds

1. Introduction

The Internet is a massive network infrastructure on a global scale that interconnects LAN- and broadband-access users through WAN. It currently supports a wide range of social, business, and individual activities and is expected to provide even more support in the years to come. A major factor in this explosive and ongoing growth of the Internet is the rapid progress of LAN and WAN. These two types of networks are significantly different in several respects. They each have their applicable area, design ideology, and applicable technology and they differ in lifetime, reliability, and other system features. Yet at the same time, they have both been supported by technological progress over the last 20 years especially in optical communications, LSI technology, and computer technology, and they both have made significant gains on the path to ultra-high-speed and large-capacity systems. Figure 1 shows this trend toward higher speeds and capacities in terms of line speeds and network-interface speeds (per wavelength) [1,2]

High-speed WAN technology in Japan has undergone a number of developments over the years. The transition from Plesiochronous Digital Hierarchy (PDH) to Synchronous Digital Hierarchy (SDH) began in 1989 (FTM-150M, FTM-600M), and a 2.4-Gbit/s system (FTM-2.4G) and 10-Gbit/s system (FA-10G) were introduced in 1991 and 1996, respectively. In more recent years, the application of Wavelength Division Multiplex (WDM) technology has taken off around the world, and with the aim of increasing the transmission speed per wavelength, standardization of a 40-Gbit/s system had been completed [3,4] and the development stage has begun [5].

Speed has likewise been increasing for Ethernet systems, which is basically synonymous with LAN systems. In the 1980s, which saw the dawn of LAN technology, speeds of 10 Mbit/s (shared media) were the norm, but the introduction of 10BASE-T star topology in 1990 and 10BASE-F optical interfaces in 1993 marked the beginning of progressively higher speeds as reflected by 100BASE-T (1995), 1000BASE-X (1998), and 10GBASE-R/X4/W (2002). This growth rate seems to have the momentum for overtaking the growth of WAN interface technology. But to be objective, Ethernet has maintained this steady growth by actively incorporating existing optical interface technologies, the key to higher speeds. For example, 100BASE-FX incorporates the FDDI standard [6], 1000BASE-X the Fiber Channel standard [7], and 10GBASE the Synchronous Optical Network (SONET) standard [8]. Of course, the applicable area of Ethernet is LAN and the short transmission distances of this field make for less stringent standards in respect to optical and electrical components. The fact that low-cost optical-interface components and modules essential to LAN can be developed quickly is another reason for the rapid growth of Ethernet.

As a result of the above trends, we are now in a historical period in which advances in WAN interface technology are crisscrossing with those in LAN interface technology. For the future, we expect to see WDM development and the convergence of interface technologies as two major trends. More specifically, the number of common technological elements shared by these two types of interfaces will increase and seamless connection between LAN and WAN systems will become possible.

Against the above background, this paper presents a survey of Inter-frame Link Signaling (ILS) technology researched and developed on the basis of LAN interface technology. The objective here is to provide new seamless interface technology for more efficient transport of LAN packets and frames. The paper is organized as follows. Section 2 introduces the ILS concept, Section 3 describes Virtual SDH on Ethernet technology for assembling virtual SDH/SONET frames and ILS coding techniques, Section 4 describes PHY-LSI technology for implementing ILS, Section 5 introduces an optical transceiver (XENPAK) mounted on the same LSI, and Section 6 presents a system experiment.

2. ILS Concept

2.1. *Approach to technology development*

Interface technology for seamless connection between LAN and WAN should have a healthy balance of interface features from both types of networks. In other words, it should feature the simplicity and low-cost nature of LAN interfaces as well as the high reliability of WAN interfaces. Here, the approach to technology development can be broadly divided into two types: an approach from WAN technology and an approach from LAN technology (Fig. 2). In the WAN approach, Packet over SONET (POS) [9] and WAN-PHY (where PHY denotes the physical layer in IEEE802.3) [8] technologies appropriate SDH/SONET frames to map LAN packets

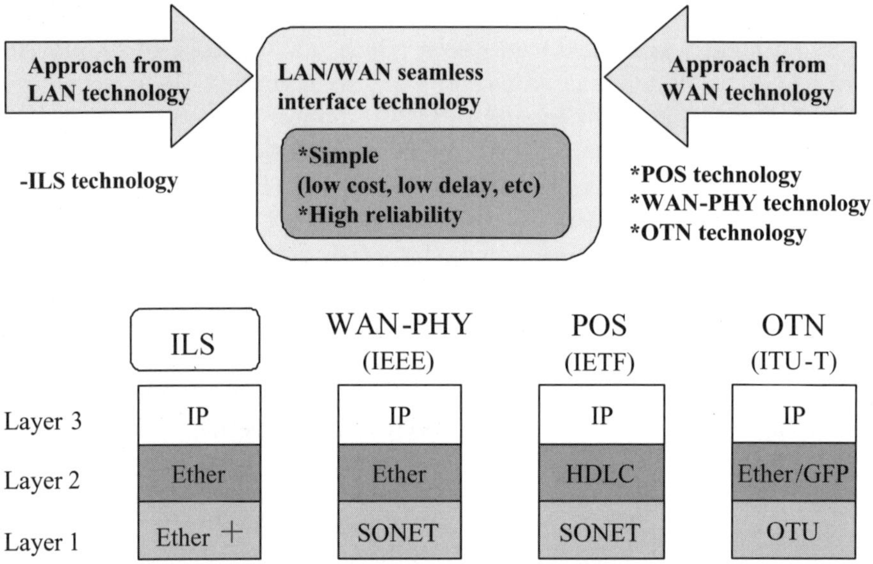

Fig. 2. ILS concept (approach and layer configuration)

or frames and perform protocol conversion. Similarly, Optical Transport Network (OTN) technology [10] performs protocol conversion by mapping LAN packets or frames to a frame structure employing Forward Error Correction (FEC) frames in undersea SDH/SONET transmission systems. In contrast, ILS technology aims to provide a WAN interface with practically no protocol conversion of ultra-high-speed Ethernet signals as an approach from LAN technology. In short, the research presented here aims to achieve a simple, low-cost, and high-reliability interface with a LAN/WAN protocol-conversion section as small and as seamless as possible.

2.2. *ILS layer configuration*

Figure 2 also shows the layer configuration in ILS together with that of competing technologies (we assume here that Layer 3 is IP in all cases). In POS, HDLC frames are used in Layer 2 and SONET frames in Layer 1, while in WAN-PHY, Ethernet frames are used in Layer 2 and SONET frames in Layer 1. In OTN, either Ethernet frames or General Frame Procedure (GFP)[11] frames are used in Layer 2 while OTU frames are used in Layer 1. Incidentally, POS conforms to the IETF standard, WAN-PHY to IEEE802.3, and OTN to ITU-T.

ILS, on the other hand, features the use of Ethernet technology in both Layer 2 and Layer 1. This implies more than just a small protocol conversion section, it says that there is no protocol conversion section and that interface conversion can be omitted for the most part. At the same time, Ethernet is a LAN standard, and WAN requirements cannot be satisfied on the basis of Ethernet Layer 1 functions alone. For this reason, ILS is a technology that adds functions to satisfy WAN requirements while making use of Ethernet Layer 1. In more general terms, ILS is a scheme that adds functions required of WAN interfaces to LAN interfaces.

A WAN interface requires system-oriented functions (or mechanisms) that make for stable and high-reliability network operation. One reason for the high level of network reliability achieved by conventional SDH/SONET WAN-interface technology is the provision of extensive maintenance and monitoring functions (OAM&P: Operation, Administration, Management, and Provisioning) for the transmission path (fiber and network nodes (repeaters, multiplexed terminals, etc.)). Specifically, the state of the transmission path is always being monitored to isolate fault locations quickly in the event of quality degradation or failures and to take appropriate maintenance actions such as issuing alarm signals. There are also protection functions for high-speed switching to redundant or reserve circuits. This is all made possible by designing a dedicated OAM&P overhead (OH) area in the SDH/SONET frame and by establishing procedures for communicating various types of maintenance and monitoring information every 125 us and for processing that information. This maintenance and monitoring information in OH can be roughly divided into frame-synchronization data, transmission-path quality (calculation and notification of bit error rate (BER)), quality degradation and failures, section/path traces, and data related to a dedicated maintenance channel. The above functions provide a means of guaranteeing WAN communication quality and achieving a high-reliability network. They are functions that ILS must be equipped with.

2.3. *ILS signaling overview*

Figure 3 shows how OAM&P information can be added to Ethernet Layer 1 using a conceptual diagram. We see here that Layer 2 includes an inter-frame gap (IFG) with no signals between Ethernet frames (Media Access Control (MAC) frames). These Layer 2 signals become Layer 1 signals by performing 64B66B or 8B/10B code conversion as a form of transmission-path coding. At this time, IFG is replaced by a special code called IDLE (or /I/).

This IDLE code period (or IFG section) is not simply an interval with no frame. It is also a location used for adjusting differences between clock frequencies so that Ethernet devices can operate independently and synchronously. This is done by inserting or removing bytes in or from the IDLE code. The standard specification here is that IFG on the send side must have at least 12 inserted bytes while a minimum of 4 bytes must exist on the receive side. The numerical values specified by this specification, however, originate in the need to maintain compatibility between MAC frames and Layer 2 (to maintain MAC-frame and Layer-2 specifications regardless of transmission speed) and are somewhat excessive as a result. In actual Ethernet systems operating at 100 Mbit/s or faster, a large number of IDLE codes exist far above the minimally required number needed for adjusting clock-frequency differences and many are not used as a result. The amount of required IDLE coding

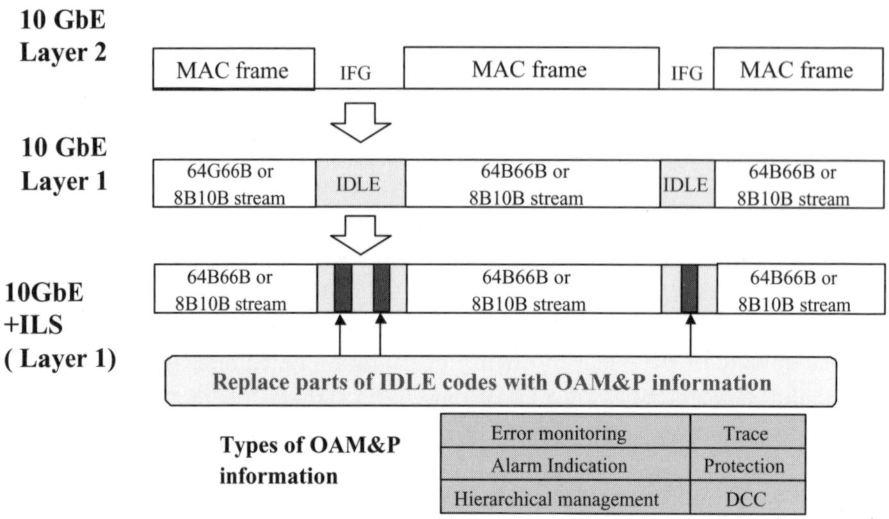

Fig. 3. ILS signaling overview

would be minimum for MAC frames having a maximum length of 1.5KB and IDLE codes having a minimum length of 12B. In a continuously flowing state, a rough calculation here reveals an IDLE-coding bit rate (with no plans for use) of several Mbit/s for GbE and several 10 Mbit/s for 10GbE. In most cases, the amount of IDLE coding is actually larger and we can consider replacing parts of IDLE codes with OAM&P information. This would enable the creation of a low-cost Layer-1 interface that is transparent to Ethernet without wasted use of Ethernet or user bands and with practically no protocol conversion. This is ILS technology.

2.4. *ILS applications*

As described above, ILS is an interface technology that adds OAM&P information to parts of IDLE codes unused by users or Ethernet devices with almost no processing of the 10GbE bit stream. From the viewpoint of constructing an optical network infrastructure, ILS can be thought of as Network Node Interface (NNI) technology for an optical network having a 10GbE User Node Interface (UNI), or for that matter, as a NNI based on Ethernet technology. In the following, we present two application examples. The first one applies ILS to a wide-area Ethernet network in the access/metro field and the second to an optical transport network (OTN) in the core-network field.

2.4.1. *Wide-area Ethernet network (application in the access/metro field)*

Recent years have seen a remarkable expansion of wide-area Ethernet services that use optical fiber to extend communication distances and construct wide-area LANs.

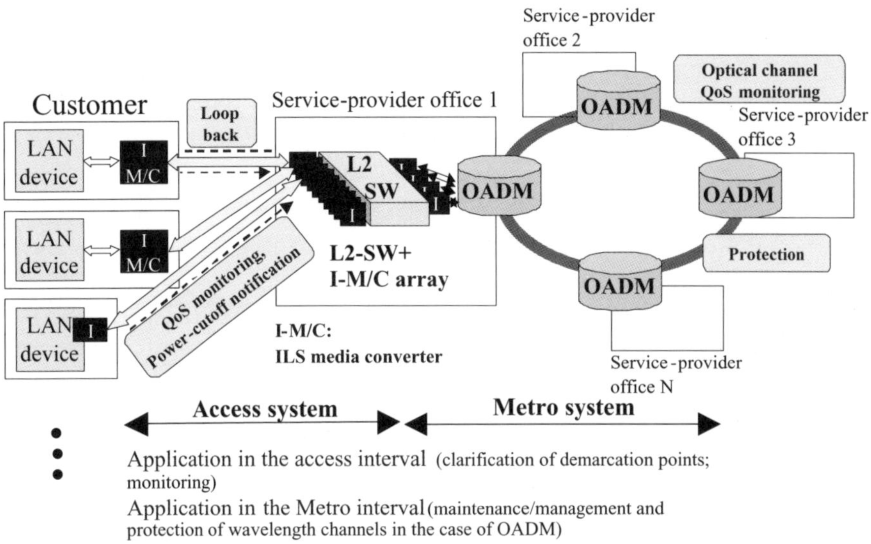

Fig. 4. application for wide-area Ethernet services

As a network infrastructure, these LANs provide Ethernet-UNI access circuits and VPN services. Here, to extend the connection distance of LAN switches, media converters are used in the access interval to convert from unshielded twisted pair cable (UTP) to optical fiber or from multi-mode optical fiber to single-mode optical fiber. And in the metro interval, a system like Coarse Wavelength Division Multiplexing (CWDM) or optical add/drop multiplexing (OADM) is used to increase bandwidth and provide system flexibility. To improve the monitoring capability and reliability of a wide-area Ethernet network of this kind, low-cost functions for maintaining and monitoring the media converters and the CWDM or OADM system must be achieved in some way. Applying ILS technology for this purpose can improve network monitoring capability and reliability. Figure 4 shows one way of configuring a wide-area Ethernet network and how ILS can be applied.

In the access interval, for example, ILS technology can be implemented in media converters installed on user premises or in transceivers installed in user LAN switches. This enables loop-back testing from the office side and remote monitoring as well as clarification of demarcation points. It also promotes a high-reliability Ethernet network through functions like transmission-path quality monitoring and power-cutoff notification. In the metro interval, ILS technology can be applied to quality monitoring and protection of each wavelength channel in the OADM ring.

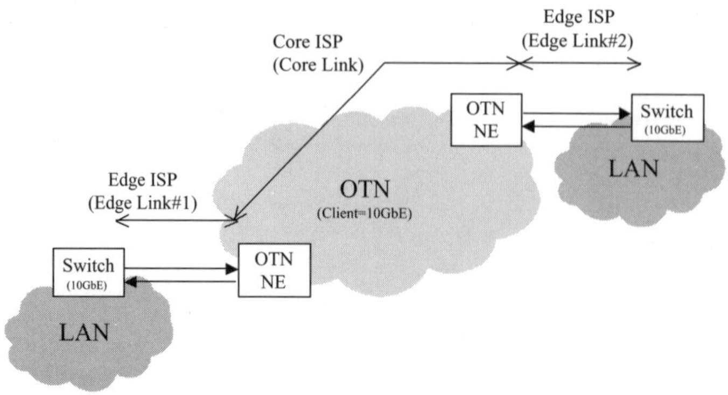

Fig. 5. Interconnection of remote LANs via OTN

2.4.2. *OTN (application in the core-network field)*

Optical transport networks based on WDM and wavelength routing have been standardized for use as WAN core networks and are expected to become widely used in the years to come. They are envisioned to act as an intermediary network for interconnecting widely dispersed LANs or wide-area Ethernet networks (Fig. 5). This format will require 10GbE as OTN clients, and maintenance and monitoring at a 10GbE PHY level will be required on three cascade-connected links as described below. Here, we can consider WAN-PHY or LAN-PHY as 10GbE PHY; Table 1 lists the advantages and disadvantages of each.

The advantages of using LAN-PHY as an OTN client in comparison to WAN-PHY are listed below.

(i) Features seamless and transparent transport
(ii) Capable of full-rate 10.3-Gb/s thereby guaranteeing the 10-Gb/s user band at the MAC frame level
(iii) Achievable at low cost with no protocol conversion
(iv) Enables use of low-cost LAN-PHY devices as LAN-PHY becomes a major format in a large LAN market.

At the same time, maintainability and monitoring capability deteriorates when using LAN-PHY as an OTN client since it has no Layer-1 OAM&P information-transfer mechanism. Referring to Fig. 5, the original mechanism was for the OTN domain and LAN domain to check the status of the LAN-PHY layer in their respective operation domains and to inform each other accordingly. Without such a mechanism, it would not be possible to isolate faults with any precision and it would be necessary to dispatch personnel adding to the cost and time of maintenance.

If ILS were to be used for transferring OAM&P information in a hierarchical manner, a fault occurring between end-to-end switches could be quickly isolated by checking, for example, edge-link 1, edge-link 2, and the core link in that order.

Table 1. Comparison of LAN-PHY and WAN-PHY features.

Items	LAN-PHY	WAN-PHY
Bit rate	○ (Mac rate 10.0Gb/s)	△ (Mac rate 9.3Gb/s)
Cost	○	△
Mainatainability and monitoring capability	×	○
Seamless connectivity	○	△

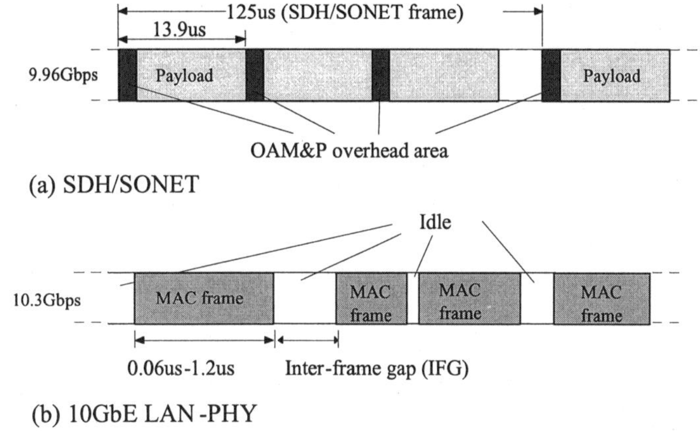

Fig. 6. Frame structure of SDH/SONET and 10GbE LAN-PHY

This approach would raise the efficiency of the network and provide faster system recoveries.

3. ILS Technology

To give a more detailed view of ILS technology, we here describe Virtual SDH/SONET on 10GbE, a technology that enables the transfer of SDH/SONET-equivalent OAM&P information, and ILS coding.

3.1. *Virtual SDH/SONET on 10GEthernet*

To begin with, the emulation of SDH/SONET OAM&P information transfer is significant for the following reasons.

(i) SDH/SONET is a system protocol equipped with sufficient maintenance and management functions for Layer 1 of the network (MAN/WAN).
(ii) SDH/SONET has already established a global presence and has a proven track record.

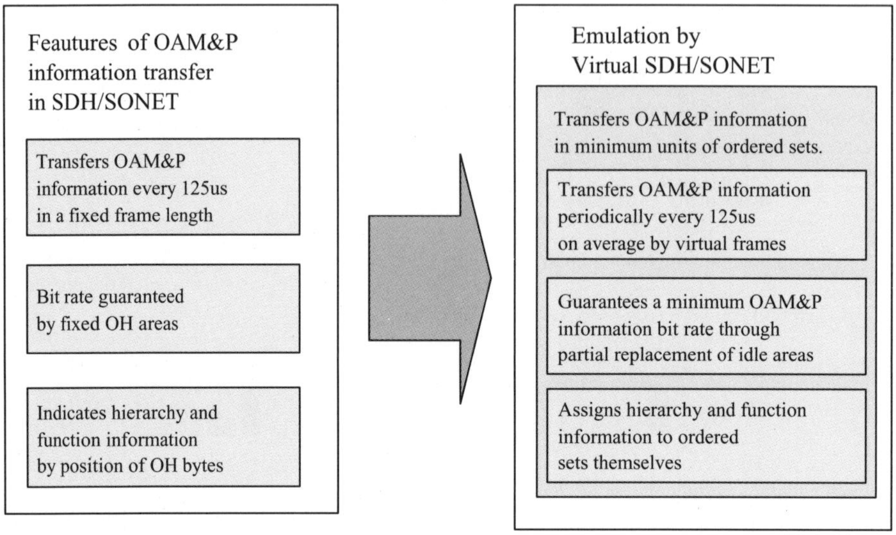

Fig. 7. Features of OAM&P information transfer in SDH/SONET

(iii) Existing OSSs for SDH/SONET can be employed and accumulated know-how can be used directly. There is therefore no need to develop a new OSS and it is relatively simple to perform a system migration. For example, a 10GbE wavelength channel can be easily added to a previously deployed SDH/SONET WDM system, or a transition from a SDH/SONET-based telephone network to an Ethernet-based data/IP network can be achieved in a smooth and low-cost fashion.

In short, emulating the transfer of SDH/SONET OAM&P information on 10Gbit Ethernet can achieve high-reliability and proven maintenance/management functions at low cost and can simplify a migration from an existing SDH/SONET system.

Next, we describe key features of SDH/SONET OAM&P information transfer targeted for emulation and present problems associated with achieving those functions on Ethernet. Figure 6 shows how the frame structures of these two systems differ [12,13] and Fig. 7 summarizes the features of OAM&P information transfer.

The transfer of OAM&P information in SDH/SONET has the following features

(i) SDH/SONET sends and receives 125-us fixed-length frames in a continuous, periodic manner and transfers various types of OAM&P information within each period.

(ii) Overhead (OH) areas for carrying OAM&P information have specific positions within each fixed frame and have a fixed bit rate.

(iii) The position of an OH area within a fixed frame dictates the hierarchy (section, line, path, etc.) and function (frame synchronization, error information, alarm, etc.) of OAM&P information to be transferred.

In comparison, Ethernet frames feature:

(i) A variable frame length from a minimum of 72 bytes to a maximum of 1526 bytes ; and
(ii) An IFG that, while having a minimum of 12 bytes as a standard specification, continues until the next frame arrives.

Ethernet frames therefore have no periodicity, and the basic idea of adding OAM&P information to idle areas as in SDH/SONET (where OH areas are given fixed, periodic positions) is consequently difficult to implement. ILS technology solves this problem by the following techniques making it possible to emulate the transfer of OAM&P information as done in SDH/SONET.

(i) Let the minimum unit for transferring OAM&P information be a 4-byte ordered set to be embedded in an idle area (IFG). An ordered set is a collection of consecutive codes used for control purposes on fiber channels (FC).
(ii) Transmit a ILS ordered set indicating a frame boundary about every 125 us and use this set to configure virtual frames. Since Ethernet frame length is variable as described above, it is not known when a frame will arrive and an exact 125-us period is not possible. The virtual frame period is therefore adjusted to be 125-us on average. In other words, the transmission time of a frame-boundary ordered set is controlled to be every 125-us on average. Other ILS ordered sets can be transmitted within this virtual-frame period so that OAM&P information can be transferred periodically.
(iii) Substitute some parts of idle areas within virtual frames by ordered sets so that an OH bit rate of about 10 Mb/s in the worst case can be obtained. This is a sufficient bit rate for transferring typical OH information (BER, DCC, trace, alarms (AIS, RDI), etc.).
(iv) Incorporate hierarchy and function information in ordered sets themselves. As described below, the 4-byte configuration of each ordered set allocates one byte for hierarchy identification ((section, line, path, etc. as in the SONET definition) and one byte for function information (error monitoring, alarm transfer, operator DCC, etc.).

Figure 8 illustrates the virtual frame concept. This scheme sends out a frame-boundary ordered set every 125-us on average to configure virtual frames and includes all types of ILS ordered sets within a virtual-frame period. On encountering an idle area, it transmits a ILS ordered set based on appropriate conditions. In addition, the reception of a frame-boundary ordered set can be used to check the correctness of the virtual frame and to perform virtual-fame synchronization and loss-of-frame (LOF) detection. The correctness of OAM&P-information transfer can

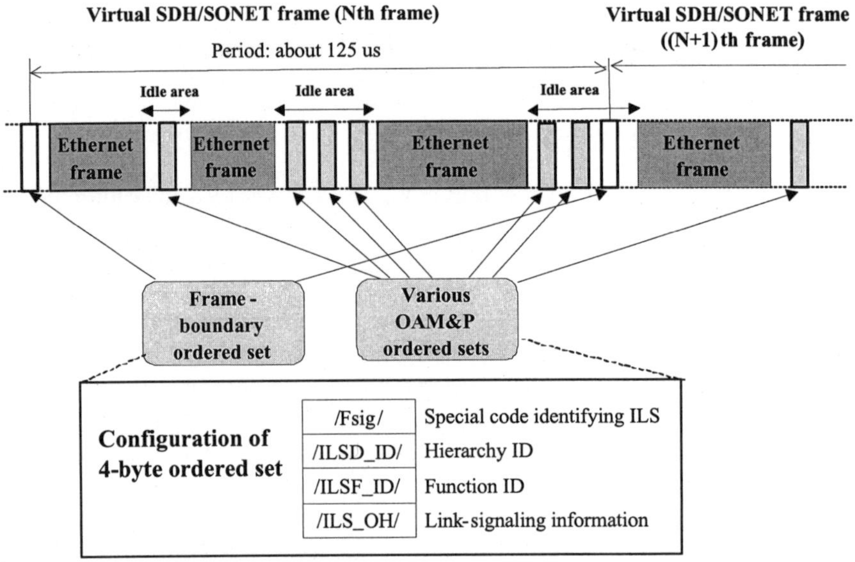

Fig. 8. Virtual SDH/SONET frame configuration

also be checked in this way. The first byte of an ordered set here is a special code indicating a ILS ordered set. The second byte holds a hierarchy ID, the third byte a function ID, and the fourth byte information equivalent to that of SDH/SONET OH or an LSM (expressing a virtual-frame boundary).

SDH/SONET defines a 3-stage hierarchy from a link-by-link section to an end-to-end path. ILS defines a hierarchy in the same way so that information can be inserted/removed or relayed by comparing the hierarch set for each device with the hierarchy ID of the ILS ordered set (Fig. 9). In the event of an end-to-end fault, this kind of hierarchical management enables immediate notification of which link the fault occurred in. ILS provides an 8-layer hierarchy.

3.2. *ILS signal coding*

3.2.1. *Requirements for ILS coding*

An issue is the format ILS should have. As mentioned above, ILS uses four-octet-wide ordered sets for transporting OAM&P information. Here, we explain why ILS uses these ordered sets and the details of the format.

Mainly, we set three requirements to determine the format.

(i) Compatibility with 10GbE LAN-PHY.

ILS should work on both kinds of LAN-PHY (10GBASE-R and 10GBASE-X), and a LAN-PHY implementing ILS should connect to a legacy one without any problem. This means that ILS should operate in a higher sublayer than

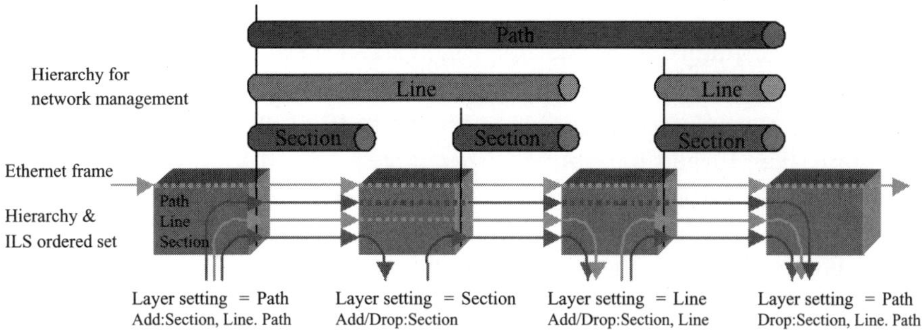

Fig. 9. Example of hierarchical management using hierarchy information

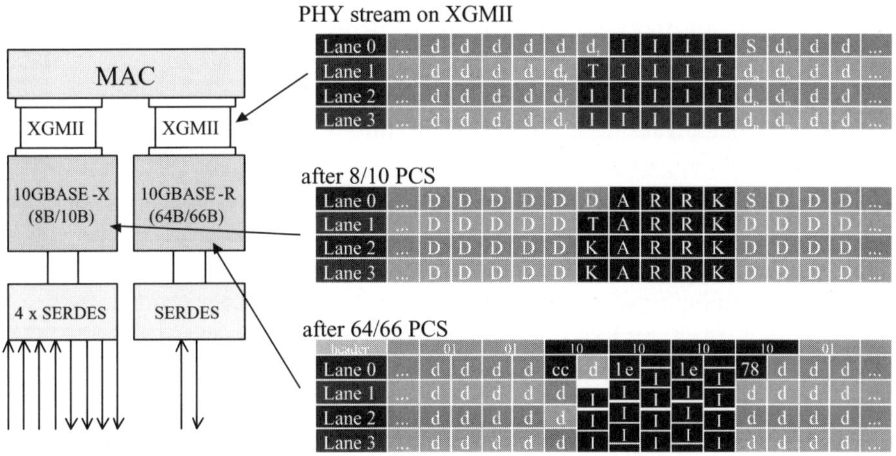

Fig. 10. 10GbE LAN-PHY sublayer diagram and data streams.

those in LAN-PHY, and ILS signal should be transparently transported through LAN-PHY.

Figure 10 shows the 10GbE LAN-PHY layer diagram and data streams. MAC (Media Access Control) frames at the data link layer are converted into a XGMII (10Gigabit Media Independent Interface) data stream by filling up Idle control characters /I/ in inter-frame gaps. XGMII is a standard interface between MAC and LAN-PHY. XGMII has four-lanes into which a data stream from the data link layer is divided. Each lane conveys one data character or control character using 8-bit-wide data and 1-bit-wide control signal.

In the PCS (Physical Coding Sublayer), the XGMII data stream is encoded by either 8B/10B (10GBASE-X) [14,15] or 64B/66B (10GBASE-R) [16]. 8B/10B is applied to each one-lane of XGMII and to serialize it, resulting in four-lane

8B/10B stream for quad-wavelength-multiplexed data transport. 64B/66B is applied two columns of XGMII for single serial data transport. These codes provide DC balance and a run length constraint in their serial data streams and enough special characters to support frame delineation and idle/error identification. The invalid data stream that is not defined in 10GbE standard is converted into error control frames in 64B/66B PCS and error control /E/ characters in 8B/10B PCS.

Compatibility means that ILS should operate in a higher sublayer like XGMII, and ILS signals should be transparently transported through XGMII, 10GBASE-X PHY, and 10GBASE-R PHY. In short, ILS signals should not be converted into error frames or /E/ characters in the PCS.

(ii) Error detection ability

ILS signal should have high integrity because it conveys OAM&P information for network management. ILS signal should have error detection ability to prevent invalid network operation caused by bit errors. In other words, it should be difficult to confuse ILS signal with the other special signals used in inter-frame gaps even though bit errors occur.

For reference, Table 2 shows the defined ordered sets and special code groups in an inter-frame gap in 10GbE. The notations of Dx.y and Kx.y are defined in Ref. [14]. The special code groups are 1-octet-wide and have corresponding fixed values. The ordered sets consist of four special characters or one special character and three data characters. The ordered sets are defined to have more than three Hamming distances from each other to maintain signal integrity in 10GbE LAN-PHY. Ordered sets are transparently transported through 10GbE LAN-PHY. Further, sequence ordered sets and signal ordered sets are reserved for future extension to convey new information by using the three data characters other than /D0.0/D0.0/D1.0/ or /D0.0/D0.0/D2.0/. The sequence ordered sets are reserved for Ethernet use, whereas signal ordered sets ||Fsig|| are reserved for transport of Fibre Channel (FC) primitive signals (ordered sets), because 10Gigabit FC (10GFC) employs almost the same PHY to convey its unique Layer-2 frame. [17,18] The signal ordered sets have the minimum Hamming distance of three against the other defined ordered sets and are transparent transported through LAN-PHY. These features are very important to consider ILS coding, as described later.

(iii) ILS signal should be disparity neutral within itself

As described later, ILS signal is based on a FC primitive signal or a signal ordered set. FC primitive signal is a four-octet-wide signal whose format is shown in Fig. 11. The first character, /Fsig/, identifies the primitive signal. /Fsig/ is the 8B/10B special character "K28.2" in 10GFC, and "K28.5" that in the legacy FC families. The remaining three are 8B/10B data characters. The three bytes identify the signaling variety, for example K28.2-D21.4-D10.2-D10.2 is the R_RDY (Receiver_Ready) function and K28.2-D21.7-VC_ID-VC_ID is the

Table 2. Defined ordered_sets and special code group.

Code	Ordered Set	Number of Code-Groups	Encoding
||I||	Idle		Substitute for XGMII Idle
||K||	Sync column	4	/K28.5/K28.5/K28.5/K28.5/
||R||	Skip column	4	/K28.0/K28.0/K28.0/K28.0/
||A||	Align column	4	/K28.3/K28.3/K28.3/K28.3/
	Encapsulation		
||S||	Start column	4	/K27.7/Dx.y/Dx.y/Dx.y/
||T||	Terminate column	4	Terminate code-group in any lane
||T0||	Terminate in Lane 0	4	/K29.7/K28.5/K28.5/K28.5/
||T1||	Terminate in Lane 1	4	/Dx.y/K29.7/K28.5/K28.5/
||T2||	Terminate in Lane 2	4	/Dx.y/Dx.y/K29.7/K28.5/
||T3||	Terminate in Lane 3	4	/Dx.y/Dx.y/Dx.y/K29.7/
	Control		
/E/	Error code-group	1	/K30.7/
	Link Status		
||Q||	Sequence ordered_set	4	/K28.4/Dx.y/Dx.y/Dx.y/
||LF||	Local Fault signal	4	/K28.4/D0.0/D0.0/D1.0/
||RF||	Remote Fault signal	4	/K28.4/D0.0/D0.0/D2.0/
||Qrsvd||	Reserved	4	||LF|| and ||RF||
	Reserved		
||Fsig||	Signal ordered_set	4	/K28.2/Dx.y/Dx.y/Dx.y/

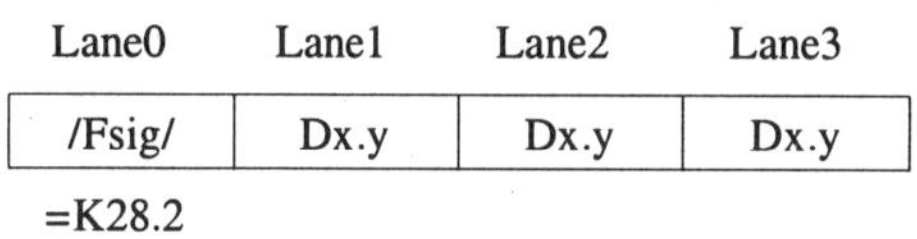

Fig. 11. Fibre Channel primitive signal.

Virtual Circuit Ready function.

Note that all FC primitive signals are disparity neutral in the legacy FC families at rates lower than 10Gb/s. The disparity means the difference in the number of ones and zeros in each character in 8B/10B coding. For example, the disparity of a 10bit character "0110010110" is 0 (neutral), while that of "0100110111" and "0100110001" is ± 2 (polarized). The disparity of each 10B character is either zero or ± 2. 8B/10B encoding requires that polarized disparity characters alternate in the character stream to maintain its DC balance. For example, R_RDY primitive signal has polarized-polarized-neutral-neutral disparity. Thus, the running disparity of the whole R_RDY primitive signal becomes neutral.

This requirement allows ILS signal to become a subset of FC primitive signals, and ILS can also work on the 10GFC as well as on the 10GbE in a more familiar manner.

3.2.2. *Approach for ILS coding*

(i) LAN-PHY compatible

We designed ILS so that it use signal ordered sets for OAM&P information transport to satisfy the requirement of transparent transport through 10GbE LAN-PHY. For transparent transport through LAN-PHY, ILS must be based on already defined signal in the 10GbE standard. There are two signals reserved for future extension to convey new information; sequence ordered sets and signal ordered sets. However, ILS ordered sets should not be based on sequence ordered sets because ILS ordered sets may overlap with newly defined sequence ordered sets for Ethernet use. On the other hand, because signal ordered sets are reserved for transport of FC primitive signals, ILS ordered sets never overlap with the other ordered sets including sequence ordered sets. Further, ILS ordered sets are defined not to overlap the already defined Fibre Channel primitive signals.

We define the format of ILS ordered set as depicted in Fig. 12. The second and third data characters identify the signaling variety, and the fourth conveys an overhead octet compatible with SDH/SONET (ILS_OH). The second character is a ILS Domain Identifier (ILSD_ID) and provides section, path, and line identification. The third character is an ILS Function Identifier (ILSF_ID) for error monitoring, tracing, protection switching, etc.

(ii) Error detection ability

One approach to achieving ILS ordered sets with error detection ability is to keep the Hamming distance from signals that could be confused with ILS ordered sets. There are several ordered sets that must not confused with ILS ordered sets in XGMII, such as Idle ordered sets, sequence ordered sets and FC primitive signals. Note that upon achieving the error detection ability in ILS ordered sets, the PCS processing must also be considered because the Hamming distance of ILS ordered sets vary before and after 64B/66B and 8B/10B encoding in PCS. This means the Hamming distance should be kept in both 8B and 10B code space.

Signal ordered sets themselves are defined to have a minimum Hamming distance of four against the other ordered sets after 64B/66B encoding by the 10GbE standard. Signal ordered sets have a minimum Hamming distance of three against the other ordered sets after 8B/10B encoding.

Defining ILS ordered sets as signal ordered sets maintains minimum Hamming distance of three against the other ordered sets. However, a bit-error protection method is still required in order to prevent confusing ILS ordered sets with other ILS ordered sets or with the legacy FC primitive signals. Thus, we use bit-error-protected characters, named the ILS character set, for both ILSD_ID and ILSF_ID.

The ILS character set is brute-force searched under the following requirements.

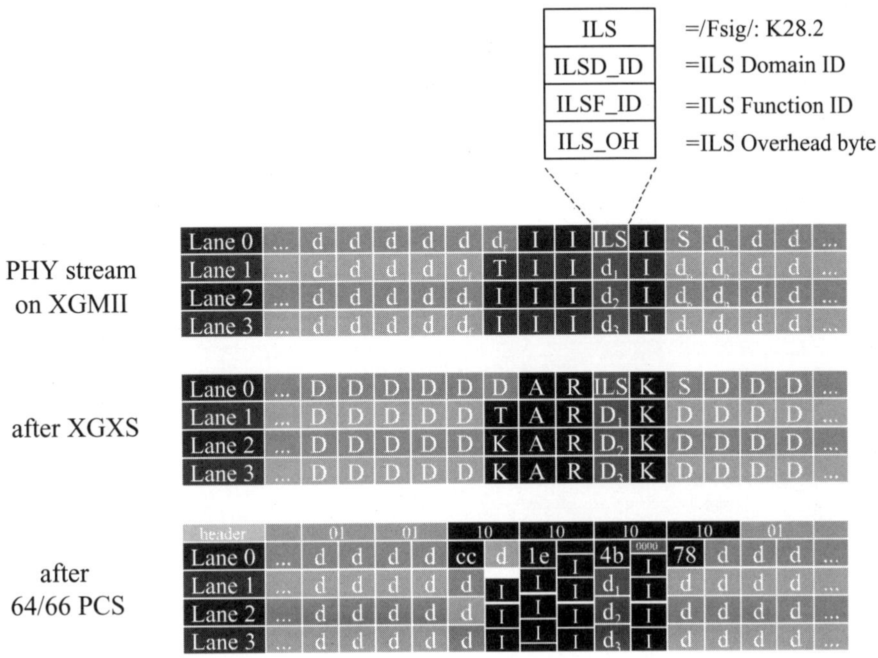

Fig. 12. ILS ordered set.

(a) Preserve the minimum Hamming distance in both 8B and 10B code spaces.
(b) Preserve the minimum Hamming distance of three in them and from the second characters of the legacy FC primitive signals.
(c) Reserve characters for future use.

The minimum Hamming distance of the ILS characters should be large in both 8B and 10B code spaces because 10GbE LAN-PHY uses 8B/10B and 64B/66B. For example, the Hamming distance of four in the 8B code space (used in 64B/66B) may become only one in the 10B code space (used in 8B/10B).

We designed the minimum Hamming distance to be three given the fact that the minimum Hamming distance is three among ordered sets in 10GbE.

To preserve the minimum Hamming distance of three from the legacy FC primitive signals, we designed the second character ILSD_ID to have the minimum Hamming distance of three from those of the legacy FC primitive signals. Neither the third character ILSF_ID nor the fourth character ILS_OH is able to have a minimum Hamming distance greater than zero, since the legacy primitive signals use these characters for octet-wide information.

Table 3 shows our search result for character sets whose minimum Hamming distance is preserved both within themselves and against the second characters of the legacy FC primitive signals. We have two character set candidates for ILS:

Table 3. Search Result.

The minimum Hamming distance		The maximum number of characters in a code			The number of possible character sets
In a character set	From the second characters of the FC-PS		Neutral	Polarized	
3	3	10	9 or 8	1 or 2	6
3	2	16	8	8	1

Table 4. Valid characters.

Neutral Disparity Name	octet value	Polarized Disparity Name	octet value
D6.6	0xc6	D18.7	0xf2
D14.1	0x2e	D1.3	0x61
D25.1	0x39	D0.5	0xa0
D25.6	0xd9	D11.7	0xeb
D19.2	0x53	D26.0	0x1a
D13.5	0xad	D3.4	0x83
D13.2	0x4d	D28.7	0xfc
D4.0	0x04	D23.1	0x37

the character sets whose Hamming distance is three both within themselves and against the second characters of the FC primitive signals, and the character set whose Hamming distance is three within itself and two against the FC primitive signals.

We adopted the latter character set because it consists of an equal number of neutral and polarized disparity characters. This allows disparity compensation by duplicate mapping in ILSF_ID as described later.

Table 4 specifies the character set prepared for ILSD_ID and ILSF_ID. It consists of sixteen valid data characters. Half of them have neutral disparity and the remainders have polarized disparity. Here "0x" means the value in hexadecimal format.

We use the eight neutral disparity characters for LSD_ID as shown in Table 5. The minimum Hamming distance is three within themselves and two against the second characters of the legacy FC primitive signal Ordered Set (D21.4, D21.7, D20.4, D5.4, D10.4, D31.2, D17.4, D31.3, D31.5, and D31.6). This enables us to detect at most two bit errors in the ILS ordered set.

Table 6 shows our assignment of two alternative characters for each ILSF_ID: a neutral and a polarized disparity character. The minimum Hamming distance is kept to three within the characters. The duplicate mapping of ILSF_ID allows disparity compensation, described later.

(iii) Disparity Neutral

Table 5. Valid characters in ILSD_ID.

Neutral Disparity Name	octet value	ILS Domain Name
D6.6	0xc6	ILS Base
D14.1	0x2e	ILS Section
D25.1	0x39	ILS Line
D25.6	0xd9	ILS Path
D19.2	0x53	reserved
D13.5	0xad	reserved
D13.2	0x4d	reserved
D4.0	0x04	reserved

Table 6. Valid characters in ILSF_ID.

Neutral Disparity		Polarized Disparity		ILS Function Name
Name	octet value	Name	octet value	
D6.6	0xc6	D18.7	0xf2	status
D14.1	0x2e	D1.3	0x61	data communication channel
D25.1	0x39	D0.5	0xa0	trace
D25.6	0xd9	D11.7	0xeb	error monitoring
D19.2	0x53	D26.0	0x1a	protection switching 1
D13.5	0xad	D3.4	0x83	protection switching 2
D13.2	0x4d	D28.7	0xfc	growth
D4.0	0x04	D23.1	0x37	order wire / signal label

In order to make the whole ILS ordered sets disparity neutral, the ILS coding process performs disparity compensation by selecting the disparity of ILSF_ID according to the arbitrary disparity of ILS_OH. As shown in Fig. 13, when the disparity of ILS_OH is neutral, the ILS coding process selects the polarized disparity character in ILSF_ID (Case 1). When the disparity of ILS_OH is polarized, it selects the neutral disparity character in ILSF_ID (Case 2). This makes the whole disparity neutral because the first character K28.2 always has polarized disparity, ILSD_ID always neutral disparity, and the sum of two polarized disparities and the sum of two neutral disparities becomes neutral.

The minimum Hamming distance from the FC Primitive Signal becomes three if only neutral disparity characters are used in ILSD_ID because of this disparity compensation. Even though the character set has the Hamming distance of two from the second characters of FC primitive signals, the Hamming distance to the ILS Ordered Set is guaranteed to increase by one by the fact that at least one bit (either the third or the fourth character) is different. In other words, the difference in the number of ones and zeros between ILSF_ID and ILS_OH characters is always non-zero, whereas that in the number of the third and the fourth characters of the FC primitive signals is always zero. The minimum Hamming distance of the ILS ordered sets from the FC primitive

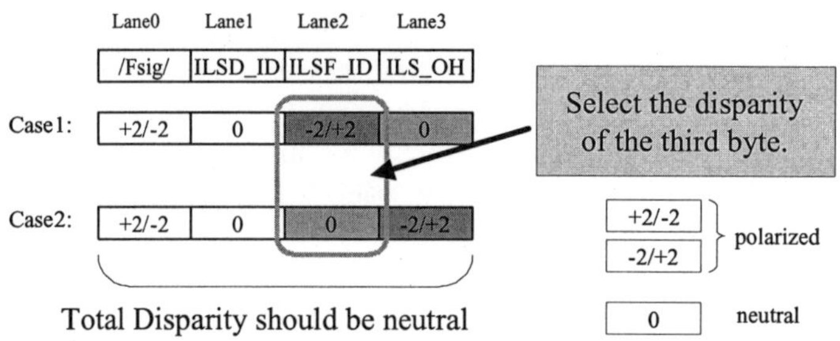

Fig. 13. Disparity compensation.

signals can be guaranteed to be three.

4. 10GbE-LANPHY LSI with ILS

ILS protocol technology has been described up to the previous section. And the LAN-PHY LSI and the optical transceiver module embedded with ILS have been developed to make it easy to implement ILS in optical transport systems or Ethernet equipment. This section described the 10GbE LAN-PHY LSI with ILS, focusing on its device and circuit technologies to achieve very high speed and low power as well as ILS implementation.

4.1. *Device technology*

We have developed 10GbE-LAN PHY chip featuring ILS function. We adopted HTI(Hybrid-Trench Isolation)-SOI(Silicon On Insulator) process technology for the PHY chip. Fig 14. shows the cross-sectional view of HTI-SOI device [23].

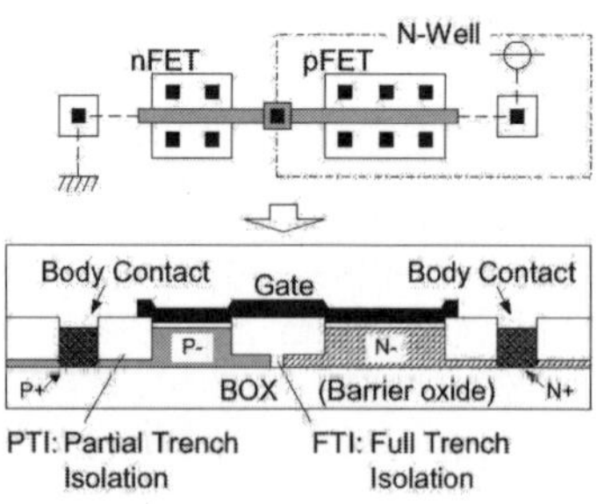

Fig. 14. Structure of HTI SOI CMOSFET.

The HTI consists of partial trench and full trench isolations. The body potential of the MOS transistors is fixed through the silicon layer under the partial trench isolation oxide. The HTI-SOI structure can thus suppress the floating body effect such as history effects, and the design layout of cell-library of bulk-device is easily translated to HTI-SOI device. Full trench isolation is applied for the boundary between nMOS and pMOS, analog circuit and digital circuit, and so on, in order to drastically suppress the latch-up and cross-talk noise. It is important for PHY chip to have endurance against latch-up because most telecommunication system is required to be hot-pluggable.

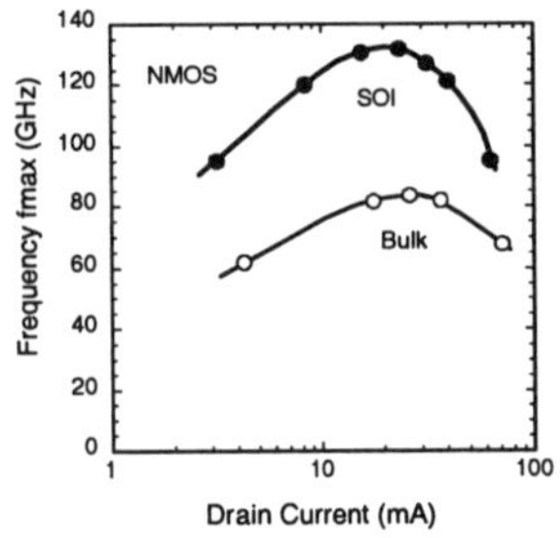

Fig. 15. Maximum oscillation frequency of the SOI and blk nMOSFET. The gate length is 70nm.

Besides, since HTI-SOI has a SOI structure, the device has a high-speed performance. Fig. 15 shows the fmax performance compared with bulk device. The fmax of HTI-SOI is 60% larger than that of bulk device [24].

4.2. *LSI architecture*

Fig.16 shows the architecture of the 10GbE LAN-PHY chip. The chip consists of PMA/XAUI SERDES, 8B10B Enc/Dec, Elastic FIFOs, 64B66B Enc/Dec and ILS blocks. The function of these blocks are discussed in this section.

(i) XAUI CDR(DES): XAUI CDR is a 4-lane CDR with deserializer. It receives the incoming data and converts them into 10-bit parallel data. The input buffer has equalizer, which will be shown later in this section.

(ii) XAUI SER: XAUI SER is a 4-lane serializer which converts 4-lane 10-bit parallel data into 4 serial data stream. The output buffer has pre-emphasis function, which will be shown later in this section.

(iii) 8B10B Encoder/Decoder: 8B10B is a block code function that encodes 8 bit data into 10 bit code, and widely adopted in Giga-bit data transfer such as GbE, XAUI etc. Since 8 bit data is mapped to a 10-bit space, there is a wide free space to have additional functions. Using this freedom, 8B10B code has merits as follows.

   (a) Can use special function in 10-bit code such as byte boundary notification (called Special Character) .

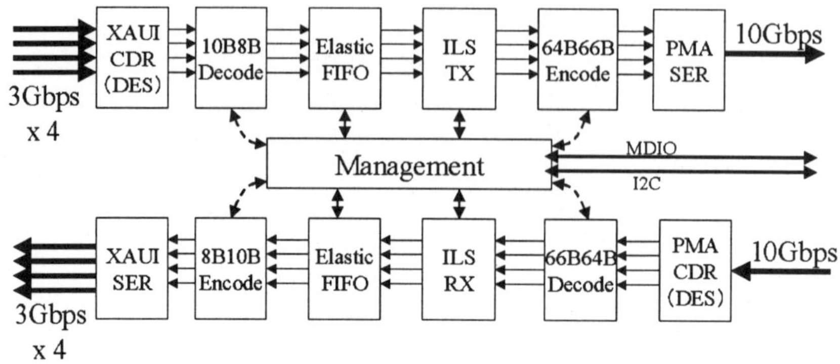

Fig. 16. Block diagram for 10GbE LAN-PHY chip.

(b) Maintains DC balance of 10-bit code and enables transmission line to be C-coupled.
(c) Limits the number of identical continuous symbols (called Run Length) to be five. This is important for CDR to ensure the quality of recovered clock.

(iv) 64B66B Encoder/Decoder: One problem with 8B10B code is its speed efficiency that needs to transfer the data 25% faster. This percentage is too much for data signaling in 10Gbps, and 64B66B code was developed in IEEE 802.3ae specification. 64B66B is a block code that codes 64-bit data into 66-bit code, just adding 2-bit sync header at the beginning of incoming data. The speed overhead is about 3%, much smaller than 8B10B. The decoder has a State-Machine to find the sync header. Unlike 8B10B function, this code doesn't limit run-length but the corresponding function is achieved by the scrambler.
(v) Elastic FIFO: In 10Gbps Ethernet specification, frequency deviation between a transmitter and a receiver is allowed if it is within +/- 100ppm. So the function to absorb the difference is needed in system. In 10GbE system, the Idle insertion/deletion in the IFG's plays this role. According to IEEE specification, there are 12-byte Idles between user payloads. PHY chip can insert/delete Idle bytes in order to absorb the frequency difference between system (XAUI side) and network (PMA side). For example, if the system side is faster than network side, IFG is inserted at FIFO between PMA-CDR and XAUI-SER (lower side of Fig. 16), and IFG is deleted in FIFO between XAUI-CDR and PMA-SER (upper side of Fig. 16).
(vi) Serial bus interface: Various information such as chip operating mode, link status, digital optical monitoring and so on is controlled/monitored via MDIO/I2C bus.
(vii) ILS(Inter-frame Link Signaling): This block manages link information via ILS

ordered sets(LS-OS) previously shown in section 3.

(viii) PMA SER: PMA SER is the 10Gbps Serializer.

(ix) PMA CDR(DES): PMA CDR is the 10Gbps CDR with deserializer function. The input buffer has DC offset canceller and equalizer to have good input sensitivity.

4.3. *Techniques for high-speed circuit*

For 10Gbps data transmission, some physical techniques are needed because system bandwidth often exceeds environmental capabilities of existing printed circuit board (PCB) technology. We introduce some techniques in 10GbE-LANPHY chip.

(i) Inductive-load buffer: At high-speed operation, the high frequency element of the 10Gbps data is likely to be distorted because of the limited bandwidth of buffer. In order to overcome this, the CML buffers which operate 10Gbps use inductors for employing the shunt-peaking effect [25](Fig.17). Figure 18 shows the effect of shunt-peaking on bandwidth enhancement in simulation. According to Ref. [25], the inductive-load extends bandwidth about 1.8 times as large as one without inductors at -3dB frequency. The effect of the bandwidth extension is most clearly seen by comparing the 1dB compression point. As shown in Fig. 18, the first stage of the inductive-load input buffer has a 1dB compression frequency about two times as large as the conventional one. Therefore by employing two stages' shunt-peak amplifiers into the input buffer, the total bandwidth of maximally flat response becomes more than two times as large as the uncompensated one.

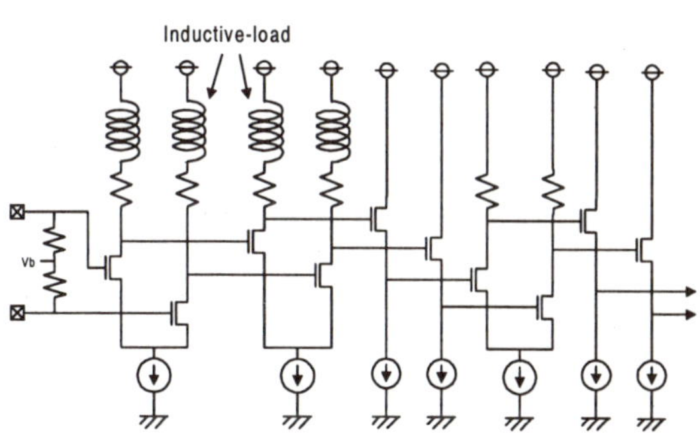

Fig. 17. Block diagram for Inductive-load CML buffer.

(ii) CDR(Clock Data Recovery): In high-speed data transmission, especially if transmission rate exceeds Gbps, eye opening at input buffer is always insufficient to capture with flip-flop cell directly. And at receiving side of the PHY chip, CDR circuit is required in order to extract the clock and re-time the in-

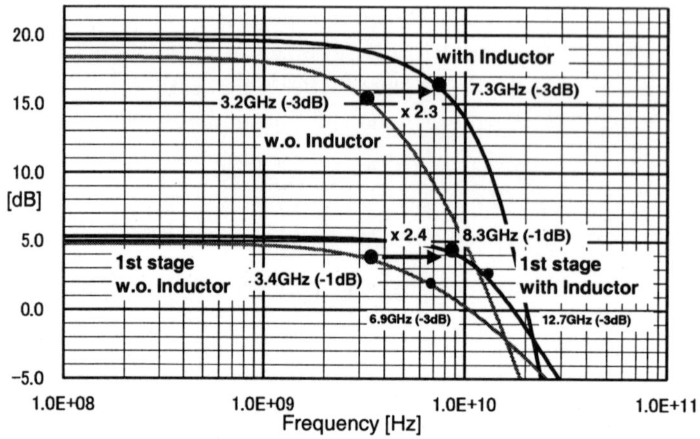

Fig. 18. Comparison between inductive-load and non-inductive-load CML buffers.

coming data at the center of the eye opening. The block diagram of the CDR is shown in Fig. 19.

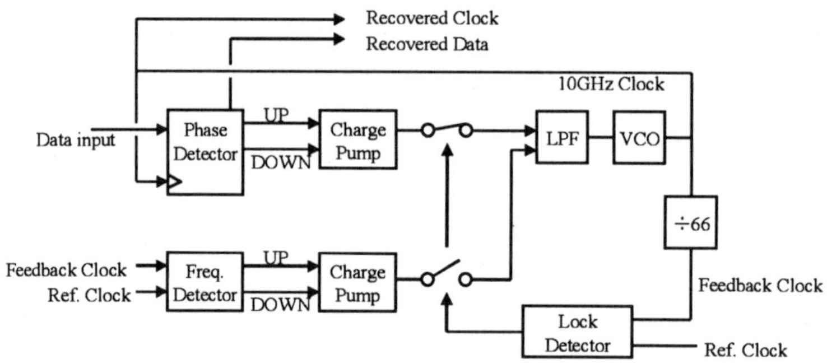

Fig. 19. Block diagram of CDR circuit.

Phase detector decides which is earlier among recovered clock and data input. Charge pump translates this early/late decision (always in digital manner) into current. VCO(Voltage Controlled Oscillator) generates clock signal, whose frequency is controlled according to the current. In this work, we adopted LC tank structure as shown is Fig. 20.

Generally, the phase noise of LC-VCO will be expressed, based on simple heuristic model, as

$$\Delta\phi \propto Q^{-2} \tag{1}$$

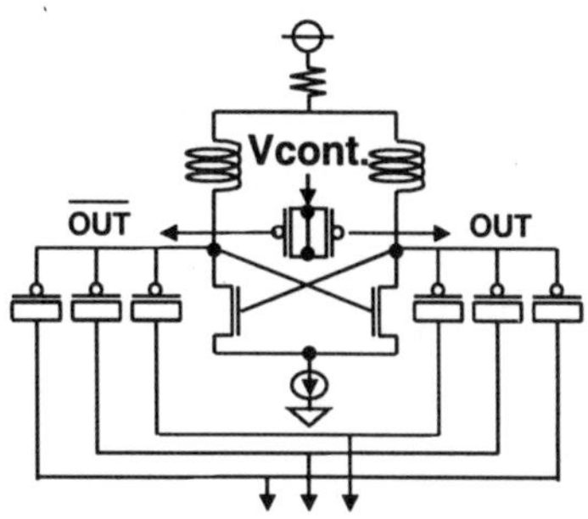

Fig. 20. LC VCO used in CDR circuit

HTI-SOI device structure makes Q factor for LC larger than bulk-CMOS device because of their high separation nature from substrate. Fig. 21 shows the measured input eye and the re-timed data output of CDR block. CDR block stably outputs recovered data/clock signal despite input eye is widely-opened or not.

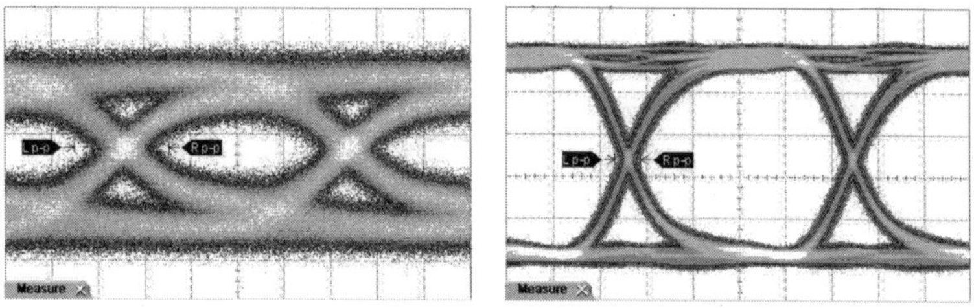

Fig. 21. Waveform of CDR output. (Left) : Data inputs. Jitter is injected through Jitter injection Module. (Right) : Data output. Total Jitter is 6.6ps p-p.

(iii) Pre emphasis(for XAUI-SER): This technique compensates the insufficient bandwidth of the PCB trace by shaping transmission waveform beforehand to boost high frequency energy. Fig. 22 shows the block diagram and timing chart of the pre-emphasis circuit. In this circuit, input signal and inverted-delayed input signal are mixed in final stage amplifier. If input signal has a data transition, the two signals work together to drive output signal. So rising/falling edge in data transition is emphasized. Fig. 23 is the actual waveforms with/without the pre-emphasis to show its benefit.

(iv) Equalizer(for XAUI DES): This technique compensates bandwidth limitation as well as Pre emphasis. Equalizer compensates degraded high frequency energy at the receiving side. Fig. 24 shows the block diagram of the equalizer circuit. By the positive feedback loop of the N+1th stage, the high frequency element of the signal is enhanced in this circuit. The time constant of this

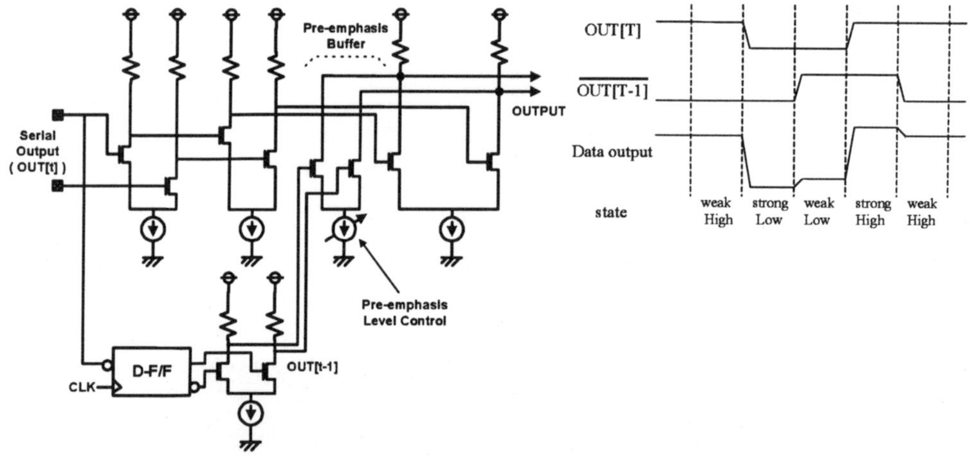

Fig. 22. Block diagram and timing chart for pre-emphasis circuit.

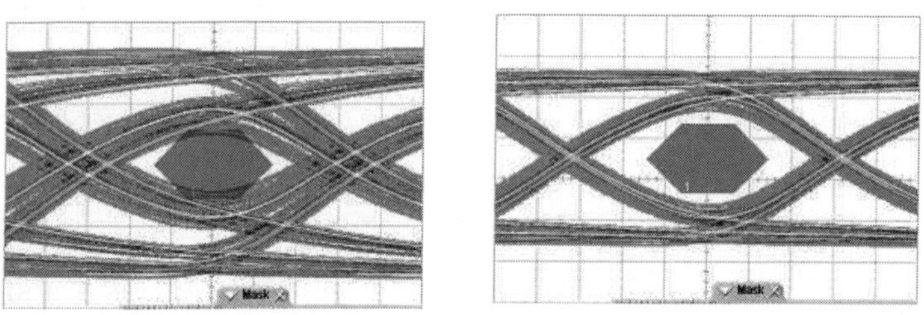

Fig. 23. Merit of pre emphasis technique. (Left) : Eye pattern without pre emphasis (Right) : Eye pattern with pre emphasis (transmission line: 700mm FR4)

feedback path is set in order to compensate the high frequency loss in the transmission line. Fig. 25 shows the simulated effect of the equalizer.

4.4. *ILS implementation*

The block diagram for ILS terminator is shown in Fig. 26. The 125us controller detects the beginning of Virtual Frame when LSb (K28.2/D6.6/D18.7/Data) packet was found. IFG detection circuit detects IFG section and enables to replace it with LS-OS. Fig. 27. shows situation of changes of data stream by ILS function. The ILS-Rx function terminates ||P|| column (ILS column) and changes into ||I|| column, ILS-Tx function vice versa. User can select for every Section/Line/Path domain whether it terminates or passes-thru. To suppress power consumption, the clock for ILS block can be gated when ILS function isn't needed, and reduces 99% of the clock speed for the portion surrounded by dotted line in Fig. 26. The HTI-SOI technology mentioned in section 4.1 enables design for this ILS block to be same

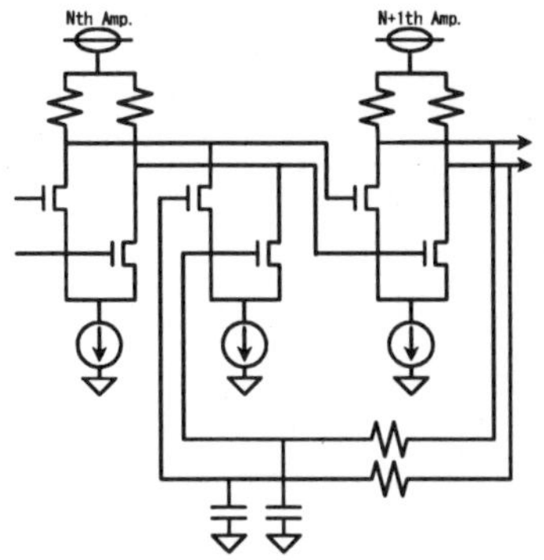

Fig. 24. Block diagram for equalizer circuit.

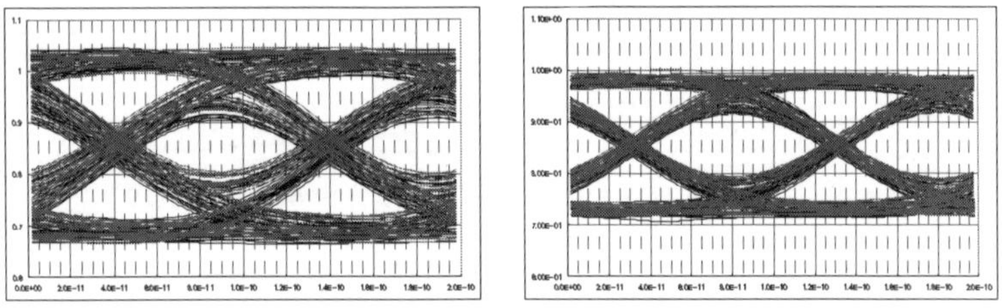

Fig. 25. Merit of equalizer circuit(simulated). (Left) : Eye diagram for transmission line without Equalizer. (Right) : Equalizer is enabled.

as bulk-CMOS manner(including CAD flow), and enables high-speed/low power consumption operation.

4.5. *LSI characteristics*

In this section, a 10GbE-LANPHY LSI with ILS was described. This chip is fabricated using 0.18um HTI-SOI process. It has XAUI/PMA transmit/receive function that meets IEEE specification, including 8B10B/64B66B encode/decode function, elastic FIFO function and ILS function. The power consumption of the chip is 1.185 W with ILS on and 1.140 W with ILS off. Thus, there is no power increase by implementing ILS. The chip size is 7.96 mm x 7.75 mm and gate count is about 220k gates (Fig. 28). Various system applications can be covered by this implementation of ILS in a one-chip PHY LSI, because a one-chip PHY LSI can be assembled in not only XENPAK[22] modules, as described in the next section, but also in other modules such as X2[23] and XPAK [24]. Furthermore, it can be also used in the combination of an XFP [25] module.

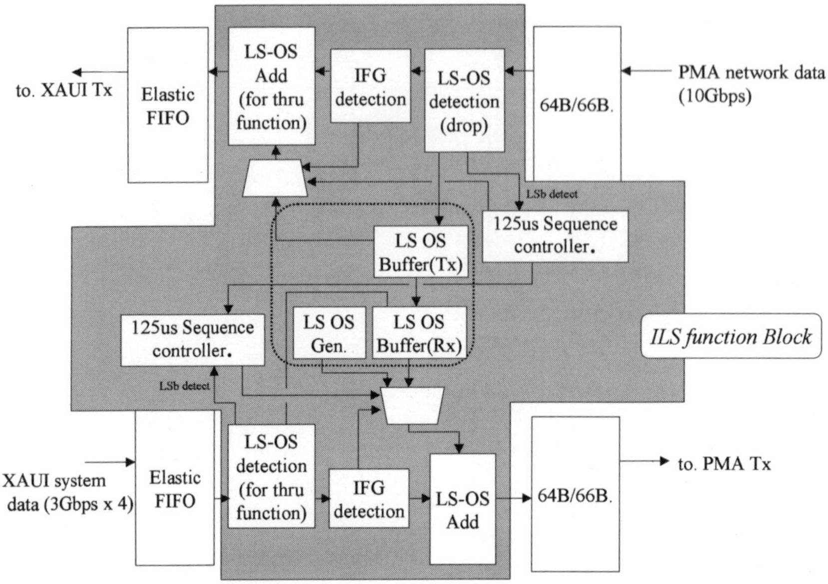

Fig. 26. Block diagram for ILS terminator

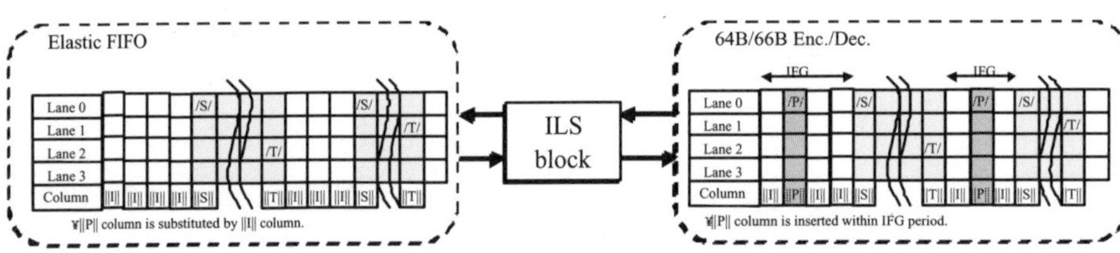

Fig. 27. Situation changes for data stream by ILS function.

5. XENPAK Transceiver

An XENPAK transceiver module with the ILS function has also been developed by using the one-chip PHY LSI described in the previous section. Since this pluggable module can cover a wide range of applications from short reach to long reach, including WDM usage in future, the ILS function can be easily implemented in optical transport systems, such as WDM systems and OADM/OXC systems, as well as in Ethernet equipment, such as L2/L3 switches and media converters or their arrays, as shown in Fig.29. In this section, the technologies to achieve XENPAK modules will be discussed.

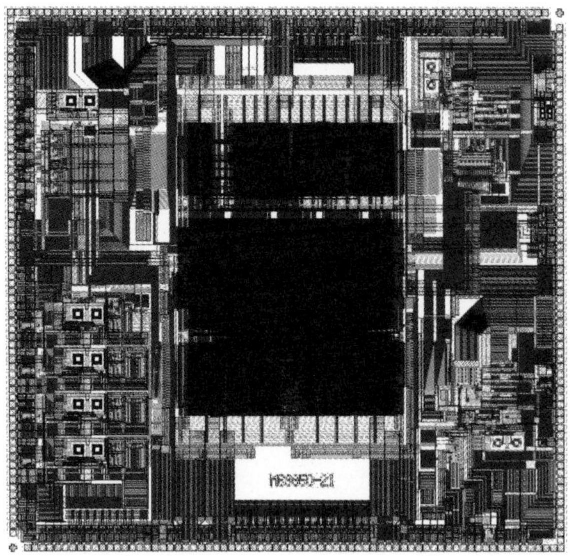

Fig. 28. Chip micrograph.

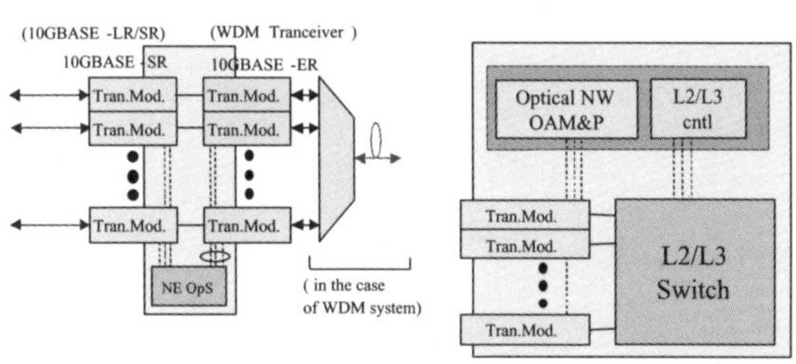

(a) Media convertor arrays (or CWDM/DWDM systems in the future) with OAM&P functions

(b) L2/L3 switches or routers with OAM&P functions of optical networks

Fig. 29. Applications of transceiver modules with ILS to transport systems and Ethernet equipment.

5.1. *General description of the XENPAK transceiver*

XENPAK is the first Multi Source Agreement (MSA) developed to realize 10-Gbps pluggable transceiver (Fig.30). The first revision of the XENPAK MSA document was publicly released in May 2001. It offers 850 nm Serial, 1310 nm WWDM, 1310 nm Serial, and 1550 nm Serial optical modules. With its thermal performance, XENPAK can support all applications in IEEE 802.3ae from SR to ER, and 80km (using APD) DWDM as extended version, even at high case temperature with enough reliability. Using dense WDM function, total throughput of 10G Ethernet

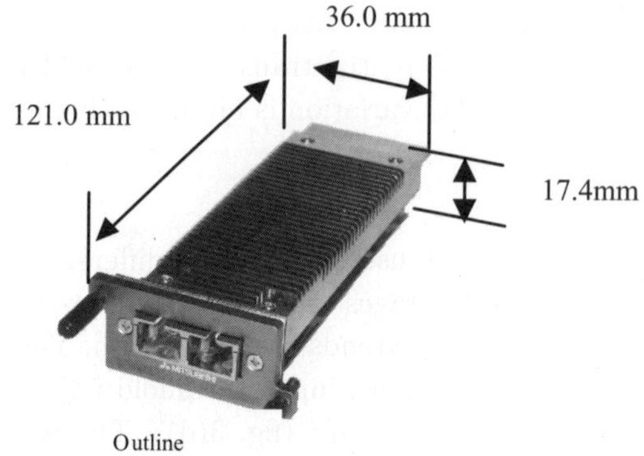

Fig. 30. XENPAK with ILS

optical port, for example, can be enhanced up to 40Gb/s or 100Gb/s more. All port types are interchangeable via front panel pluggability and XENPAK enables "Pay-as-You-Populate" cost structure during installation.

To implement the ILS function, we choose XENPAK as a platform because of its ability to line up all variations of transceivers from short to long reaches. It is also important that the XENPAK MSA specifies the register address space for the LSS (previous name of the ILS) function.

Hereafter, components for XENPAK transceivers and its evaluation results are described.

5.2. *Block diagram of XENPAK transceiver*

Figure 31 depicts the block diagram of the XENPAK transceiver. It contains PHY LSI, Laser diode, LD driver, photo diode and pre amplifier. It also has DOM (Digital Optical Monitoring) function to monitor the health of the XENPAK transceiver. The DOM function is defined in XENPAK MSA, and it monitors such parameters as temperature, laser bias currents, receive optical signal levels, and PHY LSI link status. Among these components, previous section discussed the PHY LSI in detail. The remaining key components are the optical devices, namely, TOSA and ROSA. A rough sketch of the optical devices is shown later in this section.

5.3. *Optical devices for transceivers*

For 10Gb/s Ethernet applications up to 10km+ transmission distance, uncooled DFB laser at the wavelength of 1.3um is the key device in transmit section. The critical performances for uncooled DFB laser are, wide operating temperature (over 80degree C), high relaxation oscillation frequency, and stable single longitudinal mode oscillation. We have developed a laser diode module [26] with a InGaAsP facet selective growth buried heterostructure DFB laser diode with 1/4 phase shifted

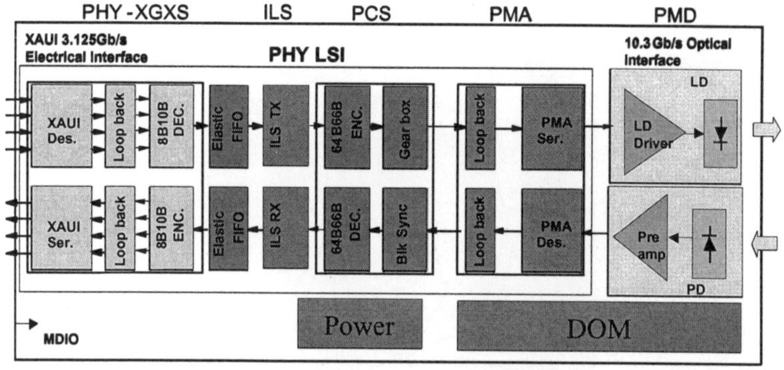

Fig. 31. XENPAK internal block diagram

grating on p-InP substrate. This laser completely eliminates side mode oscillation in success. The photograph of this module is shown in Fig. 32, and the output optical waveform is shown in Fig. 33 with Ethernet mask using fourth Bessel filter on the condition of NRZ, PRBS $2^{31} - 1$.

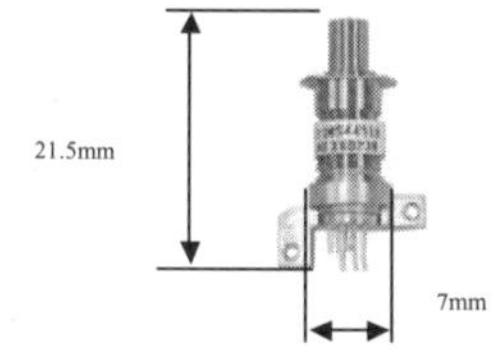

Fig. 32. Coaxial 1.3m DFB Laser module

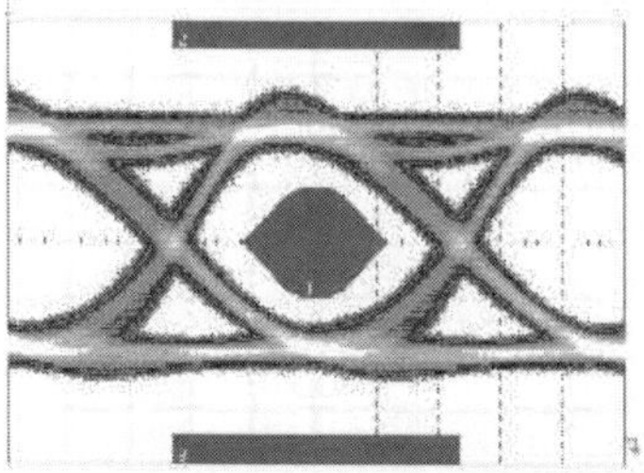

Fig. 33. Optical waveform at 80degreeC

This laser enables highly reliable transmission performance of the transceiver in any operation condition. For the next generation, AlGaInAs lasers can be used to improve the performance at high temperature.

Fig. 34. A 1.55 EML Driver integrated module

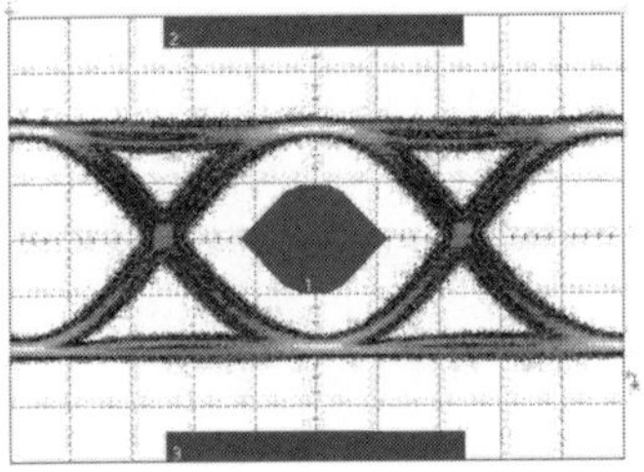

Fig. 35. Output optical waveform

For longer reach 10Gb/s Ethernet applications up to 40km+ transmission distance, Electro absorption Modulator integrated Laser, EML, is the another key device. In the past three years, 10Gb/s EML has been produced for 40km SDH market and the field performance has proven that the EML is a useful and reliable device. 10Gb/s EA is now possible to reach 80km for SONET application LR2. This 80km development is expected to contribute the cost reduction of 40km EA for 10G Ethernet ER application.

For ER application, another key parameter is the power consumption for thermoelectric cooler. Due to new assembly thermal design and using improved thermoelectric cooler, we realized small size EML driver integrated module. Figure 34 is the photograph of this EML module, and the output waveform is shown in Fig. 35. The other important parameter for the 40-km operation is the transmit penalty, and this performance is shown in Fig. 36, which is measured using our XENPAK transceiver with ILS function.

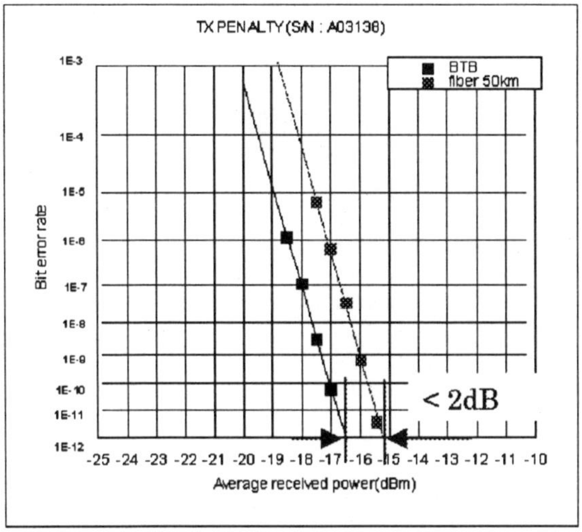

Fig. 36. 40km power penalty test results

For receiving section, we are using PD with GaAs HEMT trans-impedance preamplifier IC and could meet ER tough sensitivity specification. This PD coaxial receiver module [27] has a catadioptic system which consists of an offset parabolic mirror and a spherical lens as depicted in Fig. 37 and Fig. 38. This optical system made a coplanar assembling possible with high optical coupling efficiency of 98% from a single mode fiber to a photo diode with $20\mu m$ diameter detectigve area.

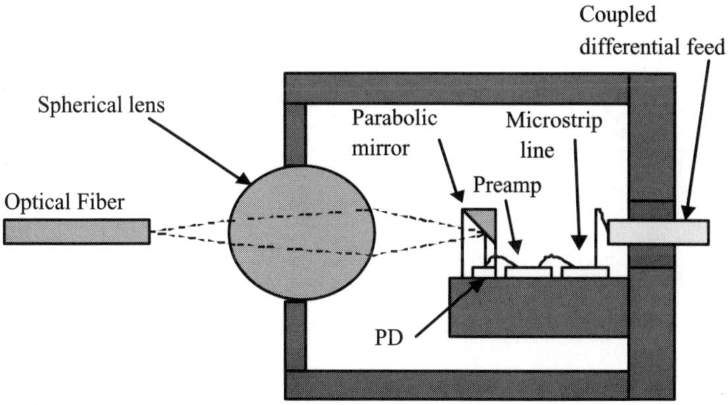

Fig. 37. Structure of PD receiver module

Fig. 38. photograph of receiver module

5.4. *XENPAK characteristics (example. 40km)*

The characteristics of XENPAK transceiver with ILS function is summarized in table 7. We could successfully meet the IEEE 802.3ae and XENPAK MSA specifications. Total power is 4.51W, well below the 6.0W XENPAK specification. Through the 2-wire MDIO control interface specified in IEEE 802.3ae, the ILS function is fully manageable. When there is an alarm in ILS function, this XENPAK module can be programmed to reflect ILS alarm to the Link Alarm Status Interrupt pin defined in the XENPAK MSA. The DCC function of the ILS is also implemented and unused pins in the XENPAK MSA are allocated for this purpose. This XENPAK transceiver is suitable for media converters, switches and routers.

Table 7. XENPAK characteristics.

No.	Test Items	Spec.	Result
1	Optical output (OMA)	<-1.7dBm	1.8dBm
2	Center Wave Length	1530 to 1565 nm	1536.9 nm
3	Extinction Ratio	>3.0dB	10.1dB
4	TX Penalty	<3.0dB	1.2dB
5	RX Sensitivity OMA	<-14.1dBm	-16.7dBm
6	Source Current	AVDD <1.5A (3.3V)	AVDD = 0.76A
		AVCC <0.5A (5.0V)	AVCC = 0.40A
7	Power Consumption	<6W	4.51W

Test Conditions: Ta=25degreeC, VDD=3.3V, VCC=5.0V, Test Pattern: PN31

6. Experiment using XENPAK Transceivers

6.1. *The benefits of ILS in the OTN network*

In the legacy LAN-PHY based network shown in Fig. 5, the OAM&P functions compatible with SDH/SONET are not provided. Suppose an edge ISP administrates an end-to-end 10GbE LAN-PHY link and an core ISP provides transparent transport service of 10GbE LAN-PHY signal. In this situation, there are several problems.

One is that both ISPs cannot always detect a defect without notification to each other. For example, when a defect occurs at link#1, Edge ISP cannot localize the fault point. The 10GbE standard PCS provides LF/RF (Local Fault/Remote Fault) signaling function. In this case, LF signals are transmitted downstream and LF is detected at Switch #2. RF signals are generated at Switch#2 and transmitted to Switch #1, and RF is detected at Switch #1. Edge ISP can know the defect occurred somewhere in the end-to-end link from link#1 to link#3 but cannot localize where the defect occurred because LF/RF signals do not contain any information other than the defect occurrence. Core ISP cannot detect a defect when it occurred in link#3 or link#4, either. This lack of OAM&P functions makes it difficult to localize the fault point alone.

ILS solves this problem using its SDH/SONET compatible OAM&P functions. The layered bidirectional alarm report functions facilitate the fault localization and make clear who has the responsibility for the fault independently. Table 8 shows relation of fault points and alarm status. In ILS and SDH/SONET, AIS (Alarm Indication Signal) is sent downstream and RDI (Remote Defect Indication) upstream when a defect occurs at each domain. For example, when a defect occurs in the link #1, Switch#1 detects RDI in the path domain and line domain. Switch#2 detects AIS in the path domain only.

Thus operators of ISPs can know the fault point immediately from the combination of AIS and RDI alarms detected at switches or transponders and can take a appropriate action for repairing the network.

6.2. *Experiment result*

We confirmed ILS functions using XENPAK transceivers in an experimental setup emulating the network shown in Fig. 5. Fig. 39 shows the experimental setup. XENPAK#1 and XENPAK#6 corresponds to Switch#1 and Switch#2, and XENPAK#2-#3 to Transponder#1, and XENPAK#4-#5 to Transponder#2. Test Frame Generator/Capture can generate/capture MAC (Media Access Control) frames. PCs can access ILS registers via the MDIO (Management Data Input/Output) interface.

We checked the alarm propagation function of ILS by making a defect at link #1 and #4 and capturing the alarm status of all the XENPAKs with PCs. We confirmed that the alarm status coincided with that shown in Table 8.

We also checked the error reporting function by making a defect at link#1 and capturing the error counts of ILS with PCs. We confirmed that the function worked correctly under the bit error rate of 4.92E-05 (under 1.0E-06). The ILS error function reports at a sufficient level because the bit error rate of 1.0E-06 is used for the threshold of the automatic protection switching in SDH/SONET.

We confirmed that the ILS functions worked correctly in the setup emulating the network shown in Fig. 39. In addition, we also confirmed that the other

Table 8. Relation between fault point and alarm status.

	Alarm Status @SW#1					@SW#2				
	Path		Line		Section	Path		Line		Section
Fault point	AIS	RDI	AIS	RDI	(AIS)	AIS	RDI	AIS	RDI	(AIS)
Edge Link#1										
from edge to core		○		○		○				
from core to edge	○		○		○		○			
Edge Link#2										
from edge to core	○						○		○	
from core to edge		○				○		○		○
Core Link										
from left to right		○				○				
from right to left	○						○			

	Alarm Status @NE#1					@NE#2				
	Path		Line		Section	Path		Line		Section
Fault point	AIS	RDI	AIS	RDI	(AIS)	AIS	RDI	AIS	RDI	(AIS)
Edge Link#1										
from edge to core			○		○					
from core to edge				○						
Edge Link#2										
from edge to core								○		○
from core to edge									○	

SDH/SONET compatible functions of ILS, such as section/path trace, remote-loopback, and DCC (Data Communication Channel), worked correctly.

7. Conclusion

This paper has described Inter-frame Ling Signaling (ILS) protocol technology and its LSI/module implementation. The ILS can realize SDH/SONET-compatible operations, administration, maintenance, and provisioning (OAM&P) functions for PHY of 10GbE networks. ILS can extend 10GbE LANs to MANs and WANs seamlessly. It adds transport OAM&P functions to Ethernet, including bit error monitoring, alarm indications, and protection switching by inserting OAM&P information in the inter-frame gap between Ethernet frames. In the protocol coding, the ILS ordered sets were designed and selected carefully and brute-forcely to achieve 10GbE PHY upper compatibility, error detection capability, and disparity neutral characteristics. A 10GbE PHY LSI with ILS function was fabricated using 0.18-um HTI-SOI CMOS technology and high-speed circuit design techniques, such as an inductive-load buffer, a high-speed clock and data recovery (CDR) circuit, and pre-emphasis & equalizer circuits, to achieve both high-speed and very low-power performance. The LSI meets IEEE 802.3ae specifications with very low-power con-

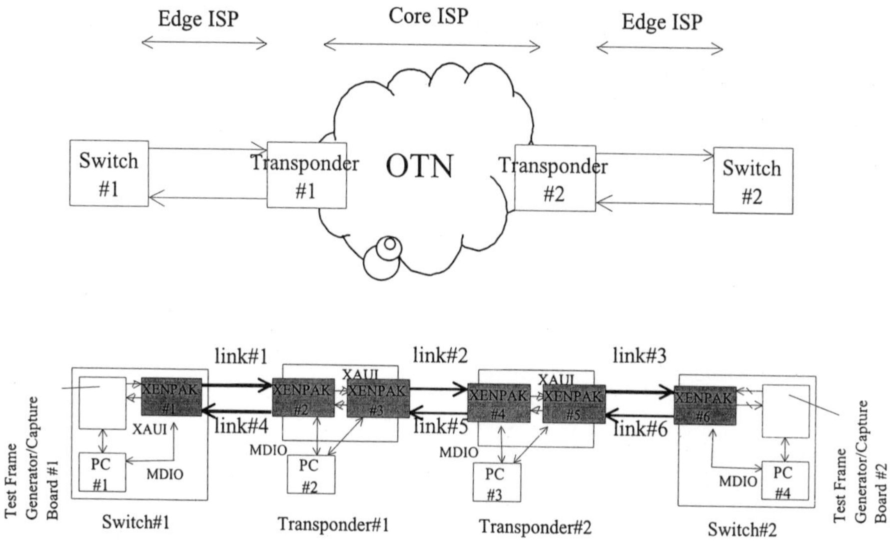

Fig. 39. Experimental setup.

sumption of about 1.2W. The power increase by the adding ILS function is less than 4% and almost negligible. An XENPAK transceiver with the ILS function using the LSI was also fabricated to implement cost-effectively the ILS function in L2/L3 switches and routers, and in the future, optical transport network systems such as WDM, OADM, OXC systems. ILS protocol operation was verified by a system experiment with XENPAK modules, which emulated the co-operation of an optical transport network and 10GbE network. The ILS protocol and the LSI/modules are expected to play an important role in achieving highly reliable and cost-effective Ethernet native WANs in the future.

References

1. M. Ohtsuka and K. Oda, "Development of very large-capacity DWDM transmission systems (in Japanese)", *NTT Technical Journal*, vol. Mar., Mar. 2001, pp. 64–66.
2. IEEE Std 802.3 2002 Edition, "Carrier sense multiple access with collision detection (CSMA/CD) access method and physical layer specifications", IEEE, 2002.
3. 40Gb/s ITU-T SDH Recommendation.
4. 40Gb/s ITU-T OTN Recommendation.
5. Y. Miyamoto, T. Kataoka, K. Yonenaga, M. Tomizawa, A. Hirano, S. Kuwahara, and Y. Tada, "WDM field trails of 43 Gbit/s/channel transport system for Optical Transport Network", *Journal of Lightwave Technol.* Vol. 20, No.12, 2115–2128, Dec, 2002
6. FDDI: ANSI X3T9.5
7. Alan F. Benner, "Fibre Channel:", McGraw-Hill,New York,ISBN 0-07-005669-2
8. N. Figueira *et al.*, "WAN Interface Sublayer (WIS) Update", http://www.ieee802.org/3/ae/public/jul00/figueira_1_0700.pdf, 2000.
9. POS; IETF RFC 2615,"PPP over SONET/SDH", http://www.ietf.org/rfc/rfc2615.txt

10. OTN; ITU-T Recommendation G.709
11. GFP; ITU-T Recommendation G.7041
12. SDH; ITU-T Recommendation G.707,or SONET; ANSI T1.105
13. IEEE Std 802.3ae 2002
14. A.X.Widmer *et al.*,"A DC-Balanced Partitioned-Block 8B/10B Transmission Code", IBM J. Res. Develop, Vol 27, No.5, pp 440–451, 1983
15. Rich Tarborek,"XAUI/XGXS Proposal", IEEE 802.3ae Task Force, Ottawa, May 23-25, 2000, http://grouper.ieee.org/groups/802/3/ae/public/ju100/taborek_2_0700.pdf
16. Rick Walker *et al.*,"64b/66b PCS", IEEE 802.3ae Task Force, Ottawa, May 23-25, 2000, http://grouper.ieee.org/groups/802/3/ae/public/ju100/walker_1_0700.pdf
17. IEEE P802.3ae draft5.0, May 1, 2002
18. 10GFC Draft 1.99, ftp://ftp.t11.org/t11/pub/fc/10gfc/01-099v4.pdf, 2001
19. XENPAK MSA: http://www.xenpak.org/
20. X2 MSA: http://www.x2msa.org/
21. XPAK MSA: http://www.xpak.org/
22. XFP MSA: http://www.xfpmsa.org/cgi-bin/home.cgi
23. Y.Hirano, Takashi Ipposhi, *et al.*,"Ultra-high-speed and low-power SOI CMOS technology with body-tied hybrid trench isolation structure", *"Dielectrics for Nanosystems: Materials Science, Processing, Reliability, and Manufacturing"*, *Proceedings of the First International Symposium Electrochemical Society Proceedings* Volume 2004-04 pp. 60–64.
24. Toshiaki Iwamatsu, Mikio Tsujiuchi, *et al.*, "Low-Noise and high-frequency 0.10um body-tied SOI-CMOS technology with high-resistivity substrate for low-power 10Gbps network LSI", Ext. Abst. SSDM, p754, 2003,
25. Tsutomu Yoshimura, Kimio Ueda, *et al.*, "A 10Gbase Ethernet transceiver (LAN PHY) in a 1.8V, 0.18um SOI/CMOS technology.", *IEICE Trans. Electron*, vol.E86-C, pp. 643–651, 2003
26. Kiyohide Sakai, Hiroshi Aruga, Shin-Ichi Tkagi, Minoru Kawamoto, Masato Negishi, Yousuke Kondoh, and Shin-Ichi Kaneko; "1.3-um Uncooled DFB Laser-Diode Module With a Coupled Differential Feed for 10-Gb/s Ethernet Applications", *IEEE Journal of Lightwave Technology*, Vol. 22, pp. 574–581, February 2004.
27. K. Sakai, M. Kawano, H. Aruga, S. Takagi, M. Negishi, Y. Kondoh, and S. Kaneko; "Catadioptic system with an Offset Parabolic Mirror and a Spherical Lens for a 10-Gb/s Coaxial Receiver Module", *ECOC-IOOC 2003 Proceedings* - Vol.3. pp. 704–705.